Sharing the Universe

Dr. Seth Shostak is an astronomer, teacher, long-time searcher for cosmic company, and science communicator. He combines this experience in a way that has made him an expert at conveying the stunning ideas behind the belief that we share the universe with other beings.

Seth is a man of diverse interests, with degrees from Princeton and Caltech. Before joining the search for aliens, Seth used large radio telescopes in both the U.S. and Holland to analyze the motions of galaxies for clues to the long-term fate of the universe. He has penned hundreds of popular articles on science and technology, and gives more than 100 talks a year on astronomical and SETI subjects.

Since 1990 Seth has held the post of Public Programs Scientist at the SETI Institute in Mountain View, California. The SETI Institute helped inspire the recent film *Contact* with Jodie Foster. As Public Programs Scientist, Seth's responsibilities include explaining to the public the fascinating work they do there, and what they are likely one day to discover. This, his first book, makes this information available to more people than ever before.

Sharing
the Universe

Perspectives on Extraterrestrial Life

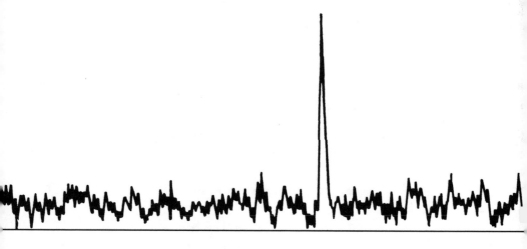

SETH SHOSTAK

BERKELEY HILLS BOOKS
BERKELEY, CALIFORNIA

Published by

BERKELEY HILLS BOOKS

P. O. Box 9877

Berkeley, California 94709

Cover and interior design by Elysium/San Francisco

Cover image of NGC 6992 courtesy of U.S. Naval Observatory

Printed by Data Reproductions/Rochester Hills, Michigan

Library of Congress Cataloging in Publication Data

Shostak, G. Seth.

    Sharing the Universe : perspectives on extraterrestrial life /

    Seth Shostak ; foreword by Frank Drake.

          p.   cm.

    Includes index

    ISBN 0-9653774-3-1 (pbk.)

    1. Life on other planets. I. Title.

QB54.S56 1998

999—DC21   97-40214

                                  CIP

Table of Contents

PREFACE

For most authors, a preface is humility's final retreat. Fronting tens of thousands of words of immodest exposition, the preface is the author's last chance to claim indebtedness to others for all that is good in his work, and take shameful, personal responsibility for all that is not.

I shall not deviate from this convention. It may be hardly original to say so, but many people have made essential contributions to the production of this book. Without their help, *Sharing the Universe* would not and could not have been written. However, despite the best efforts of others, errors of fact or interpretation may remain. They are of my own doing.

Rob Dobbin and John Strohmeier of Berkeley Hills Books lavished encouragement and expertise on this project from its earliest beginnings, an informal discussion in a Mountain View salad bar. They have been steadfast in their support and tolerance. I would also like to thank Carol Oliver for her careful reading of the earliest chapters, and Lisa Schulz for her elegant design work.

The entirety of the book has been read and commented upon by two other individuals. Tom Pierson was originally asked to review Chapter 8. Having done so, he then graciously volunteered to do the same for the rest of the book. His expert advice was invaluable. Karen Shostak scrutinized every word of this small tome several times, and blessed it with her considerable editorial expertise. Although her enthusiasm and help were unflagging, she was also willing to gently share the news when prose was tortured beyond the bounds of the Geneva Convention.

My colleagues at the SETI Institute have been both supportive and helpful, and I would especially like to thank Jill Tarter and Frank Drake. Finally, several people who are expert in fields that are foreign to me were kind enough to offer suggestions and guidance, and in some cases make major contributions to the ideas presented here. I thank Drs. L. Marino, W. A. Bonner, and L. A. Fette for both help and motivation.

Parts of Chapter 9 were originally printed as articles in *Astronomy* and *Mercury* magazines.

SETH SHOSTAK, MOUNTAIN VIEW, CALIFORNIA

The world is awash in tales and images of extraterrestrials. And why not? There is no better grist for the special effects mill, for the hucksters, and for those who would exploit our legitimate quest for adventure. Our space program and the growing prowess of our telescopes have provoked widespread appreciation of the grandeur of the cosmos. And for every new discovery, a new cosmic puzzle is recognized. We indeed live in a vast, mysterious universe rich in phenomena beyond our wildest imaginations, even now. We live in a time when great discoveries come at a prodigious pace. At the same time, the means of presenting these phenomena, either accurately or as spectacle, have become themselves one of the wonders of our civilization. The marriage of discovery and media skill has enriched us enormously, as it has educated us.

That same marriage has served well those who would exploit our fantasies, superstitions, and gullibility. People are enthralled by tales of UFOs, government conspiracies at a cosmic level, abductions of humans to magic places beyond our capabilities. Each person has a reason for his or her interest. To some, this offers escape from a humdrum existence; to others it may offer a substitute for traditional religion whose underpinnings seem suspect. Those who would exploit us know well how to ply their trade. Mix in some true observations with some fantasy, and the naive will conclude that all is true, that there is no fantasy. A false credibility is created. Books sell. People blindly join and adhere to cults. Wildly impossible science fiction movies ring true. And many people get rich.

One could argue, 'So what? If everyone involved is happy, what is the problem?' The problem is that no civilization can thrive on falsehood. In the end, false 'knowledge' leads to wrong decisions, wrong choices of technologies, a wrong distribution of resources, wrong priorities, wrong choice of leaders. And civilization as a whole is the loser. A prime illustration of this is the distribution of resources invested in attempts to understand life in the universe. There is widespread public confusion as to the relative promise of pseudo-scientific studies of UFOs and similar ilk, as compared with true scientific programs to find life on other planets, or radio signals from the creatures of distant stars. All knowledgeable scientists strongly favor the latter, but the world at large is unclear about this. The consequence is that far more attention is given to the pseudoscience than to the real science.

Perhaps the saddest aspect of this is that there is so much promise in the real science. We live in a universe teeming with exciting phenomena, some only recently discovered. Complex life exists in the dark deep of the ocean, thriving on chemical soup streaming from natural spigots in the rocks. There is life in abundance deep beneath the surface of the Earth, totally unaware of the great Sun so central to our existence. Indeed, this life needs not the Sun at all. Surely these scenarios occur many places in space. Among these potential worlds is Jupiter's moon Europa, which harbors a vast, ice-protected ocean where underwater spouts like those of the Earth's seas may be nourishing life. We live in a galaxy full of planetary systems, as we have only recently discovered. Perhaps these planets or their brethren host life, or life may possibly exist on their satellites. Of equal importance, we have recently constructed an armada of powerful astronomical instrumentation on the ground and in space that can witness all these phenomena. We have built instruments that have the ability to detect life similar to human life at huge distances. We live in a unique time, when the answers to some of the most profound questions of science, human philosophy, and even religion are within our grasp. But we must get our priorities straight.

This task can be carried out only by someone who has lived enmeshed in the hurricane-like developments in the search for life in the universe. Someone who has the wisdom to separate fact from clever fantasy. Someone who can put aside personal prejudices and produce a fair and balanced account of the key aspects of this great enterprise. Only a few qualify, and Seth Shostak is one of them. In this book is both a complete and fair picture of where we have come in our voyage of exploration into the cosmos in search of life, even intelligent life. In so doing, we are seeking not just to live one of the greatest adventures left to us. We also seek to glimpse the possible futures from which we must choose, for better or worse. As you read this book, remember that you are reading the prologue to both humanity's finest discovery, and the finest moment of decision.

Aliens Are Everywhere

Somebody should be selling tourist maps of the planet. Books with titles like "Earth on Five Galactic Cruceros a Day." Our humble world, an insignificant pin-prick in a vast, churning universe, is suddenly Tourist Destination #1 for aliens.

Most of this alien invasion plays out on television or at the local cinema. Couch potatoes watch with acceptance and remarkable calm as boob-tube visitors from space regularly mingle with the neighbors, mess with our history, or incinerate our cities. In *Third Rock from the Sun,* anthropomorphic aliens hope to assimilate our twentieth-century lifestyle, while in *Dark Skies,* the little guys from space try to change it. Agents Mulder and Scully, of *The X-Files,* are driven to expose what a paranoid government insists on covering up: the fact that we have cosmic visitors. "The truth is out there," the agents claim. But not too far out. The secretive extraterrestrials have bothered to fall to Earth.

At the movies, the roles once played by evil Nazis, stealthy communists, or rampaging Indians have all been given to space aliens. Hollywood is practicing job exportation on a massive scale. So far, these extraterrestrial extras have been severely typecast, but they're not all bad. Among the good guys are huggable, little *E.T.,* or the bemused, helpful creatures of *Cocoon* and *The Abyss.* Likewise the smooth-skinned, nicely bleached extraterrestrials who wheel into the neighborhood to make *Close Encounters of the Third Kind.* The alien white hats have traversed

1

countless light-years to amuse the kids, invigorate the elderly, or simply share a couple of meaningful moments.

Occasionally, the movie aliens will play a more enigmatic role. In *2001* and *Contact,* they manage to remain off-stage, while making their presence known by offering a few humans some perplexing, if visually interesting, trips through time and space. It seems that these cryptic critters are trying to better us, not as individuals, but as a species.

And then there are the galactic black hats, the evil aliens. This wretched refuse comes to our shores only to cause trouble. The double-jointed extraterrestrials of *The Arrival* hanker to take over the world. The smelly aliens of *Independence Day* just want to exploit it. The red planet makes one of its perennial assaults on Earth when *Mars Attacks,* and a few good *Men in Black* campaign against obnoxious, insectoid extraterrestrial scum. Hollywood aliens don't want to be taken to our leader. They want to *be* our leader. In the movies, somebody has to play the heavy, and aliens increasingly win the part.

Many individuals are convinced that these uninvited guests are not limited to works of fiction. Every year, thousands of people report seeing UFOs (Unidentified Flying Objects), and quite a few believe that these mysterious lights in the sky are alien spacecraft. In England, "cerealogists" devote their energies to studying enigmatic patterns that suddenly appear in fields of wheat. They claim these are messages from extraterrestrials who've carved the grain graphics for our benefit and edification. In America, aviators solemnly swear that the UFOs are wreaking havoc in the skies by occasionally buzzing our military aircraft.

On rare occasions, the putative aliens, although nimble enough to play aerial chicken with fighter jets and trace delicate designs in crops, make a serious navigational error. The most celebrated instance took place in the summer of 1947, when a celestial craft with a handful of diminutive crew members purportedly crashed in the desert near Roswell, New Mexico. According to popular belief, the U.S. Air Force efficiently collected both bodies and debris, and has kept evidence of the mishap secret for half a century.

The aliens don't limit themselves to merely cruising the stratosphere and defacing British agriculture, however. Some interstellar pilots climb out of their cockpits to cause trouble. The media are chock-a-block with the complaints of just-plain folk who say that pushy extraterrestrials have forced them aboard their saucers, attempted to get them pregnant, or simply removed ova or sperm for unspecified and unapproved use elsewhere.

Although a majority of the American public is convinced that aliens are making house calls to planet Earth, most scientists aren't. It's not that the researchers are unhappy with the idea of extraterrestrials. Many believe it's quite likely that other intelligent beings populate the cosmos. But the available evidence has failed to convince scientists that the extraterrestrials have either strafed someone's back yard or landed in it. Hollywood recognizes this scientific skepticism, and accommodates it by offering stories set, not on Earth, but on the space-dwellers' home turf. The *Starship Enterprise* is on a constant prowl of the Galaxy, seeking out new life and hoping it speaks English. Occasionally, the ship's captain will chat up a female alien, apparently unconcerned by the fact that such beings will have motives and plumbing radically different from Earth women. Other astronauts, such as the hapless crew of the spacecraft *Nostromo* in the movie *Alien,* encounter hungry, toothy critters who first use humans for incubation and then for nourishment. In *Contact,* Ellie Arroway puts on her headphones and hears an extraterrestrial signal that sounds like a pile driver pummeling a pod of whales. Before long, she's journeying across the Galaxy to have a close encounter of a weird kind.

It is a dizzying amalgam, a potpourri of tabloid journalism, high-budget films, and serious folk who think they've seen something otherworldly. Fascination with extraterrestrial life has reached a truly unprecedented level. Is there some reason for this?

A FLURRY OF ALIENS

Despite the new interest in aliens, the idea of cosmic company is old. The ancient Greeks suspected that Earth's creation was not special, not a one-time-only event. They proposed the existence of countless other worlds, in various stages of evolution, and populated by strange peoples and beasts. However, despite their belief in celestial neighbors, the Greeks didn't expect these unseen creatures to actually show up on the doorstep. It wasn't until the age of modern rocketry that serious claims of alien visitation became common. This is hardly surprising, of course. Once we had the tools to venture into space, the idea that others might be cruising the interstellar deeps seemed more plausible. In addition, World War II and the Cold War conditioned us to watch the skies just as improved technology was filling them with strange craft.

Since 1947, when pilot Kenneth Arnold described the first flying "saucers," the popular media have reported thousands of strange lights in the heavens and puzzling encounters between aliens and earthlings. The public finds it difficult to believe that all of this is nonsense, and is occasionally reinforced in this view by disingenuous disclaimers from government officials. Routine explanations of sightings as natural phenomena, balloons, or aircraft are clearly less interesting (and therefore, less frequently reported) than claims that the lights in the sky are due to short, hairless gray guys from another star system.

Alien beings are interesting in the same way that Bigfoot or the Abominable Snowman are interesting. But since they are other-worldly, the aliens have a utility that the lonely sasquatch and yeti don't. The aliens are not evolutionary throwbacks, condemned to haunt the cold periphery of our planet. They are creatures with knowledge and abilities far beyond our own. As humans, we have a need for something outside ourselves, something with the power to influence our lives and possibly make them better. Traditionally, God has fulfilled that need. But today, some folks think that extraterrestrials can step into God's shoes, at least when it comes to shaping our lives. In a high-tech age, aliens are seen as high-tech angels. Additionally, a deity's existence can never be proved. But in principle, the existence of aliens can; there is less demand for faith.

By this logic, having influential extraterrestrials in the neighborhood offers potential benefits. And newspaper columnists suggest a fuzzy reason why they'd want to visit now: the arrival of a new millennium. A once-in-ten-centuries event must surely have cosmic consequences.

So a lot of folks think that this big tick of the digital clock will encourage visits, although they seldom make clear why aliens and Christians should have the same calendar.

Even aside from the odd thought that extraterrestrials would tour our planet just because the date is auspicious, one might ask why, in a galaxy of a few hundred billion stars, the aliens are so intent on coming to Earth at all. It would be as if every vertebrate in North America somehow felt drawn to a particular house in Peoria, Illinois. Are we really that interesting?

It's doubtful that we are, but postulating an alien pilgrimage to our planet validates our importance; even more so if these celestial beings bother to abduct us for sexual experiment or give us a joy ride in their saucers. In light of the fact that any real aliens capable of travel from a distant star system would be enormously more advanced than us, this postulated fascination with humans seems questionable. After all, when Charles Darwin landed in the Galapagos Islands, he didn't offer the iguanas rides on his sailing ship, let alone try to impregnate them. The fact that both Hollywood aliens and those that reportedly flit about the stratosphere *do* take an interest in people suggests that they have been invented for our own purposes. Their task is to show that we are important in a vast and indifferent universe.

In addition to their philosophical usefulness, the extraterrestrials serve a critical need of the entertainment industry. When the iron curtain vaporized, so did a whole class of handy bad guys. Suddenly, Hollywood needed new threats. Movie moguls have tried to fill the breach with terrorists from the Middle East, but these antagonists are frequently crippled by the region's muddled politics and disparate groupings. Aliens, on the other hand, can be fashioned of unalloyed nastiness and can be bug-ugly in appearance. They can be maligned without mercy since, lacking a local constituency, it is never politically incorrect to call them names. There are practical advantages to letting extraterrestrials play the bad guys as well. The aliens don't demand speaking parts, are casual about wardrobe (often to the point of appearing without clothes at all), and are willing to accept robot-like roles. Whole hordes of them can be fashioned in the computer, a

5

casting possibility that didn't exist a decade ago. If budgets are really tight, the aliens can assume the form of humans (as they did in *Species* and *Contact*).

The rise in UFO sightings, the approach of the millennium, our hunger to feel important, and Hollywood's need for new antagonists are probably sufficient to explain the contemporary fascination with aliens. But the tidal wave of fictional critters has been swollen by scientific discoveries that make the existence of *real* extraterrestrials more probable. In the last decade of the twentieth century, we have finally succeeded in detecting planets around other stars. Most of these new worlds are oversized and inhospitable, it's true. That's because the techniques used to discover them are attuned to large, closely-orbiting planets. Lightweight worlds are still beyond the reach of our instruments. But where there are big planets, there are probably also small ones. It is only reasonable to assume that bantam-weight worlds, planets the size of Earth, accompany many stars, perhaps *most* stars. What was once only conjecture is now sliding into fact: there are countless worlds where alien life could begin and evolve.

Of course, a plethora of planets might not be enough to guarantee an abundance of aliens. Even though billions of earth-like worlds could pepper our galaxy, is there reason to think that any of them have fomented biology? Life is complex and subtle. But it might not be improbable. In 1996, a team of NASA and university researchers announced that they had found evidence of fossil life in a Martian meteorite. Electron microscope photos of the meteorite's innards show tantalizing, worm-like forms. Could these be the preserved remains of microbial life that once floated in ancient Martian seas, billions of years ago? The jury is still out. Paleobiologists are busy tearing into the meteorite, trying to find convincing evidence that the tiny features are long-dead germs from the red planet. But the very suggestion that Mars might have spawned life implies that biology is a very frequent occurrence. Life might be common. This profound thought, despite the tentative nature of the evidence, ensured that the Martian meteorite research was the number one science story of 1996.

Planets aplenty, and hints that biology is commonplace: when it comes to intelligent extraterrestrials, the musings of fiction are being given shadowy substance by science. That's what sets them apart from other imagined beings. Aliens, like ghosts, gremlins, angels and the devil, all appear in film and television. But no one seriously expects that ghosts, for instance, will prove to be real. Ghosts will never become a subject of mainstream science with their own, refereed journal, feature stories in *National Geographic,* and displays in the world's museums. The aliens are in a different category. There is a chance that tonight, next week, or next year, an experiment will prove the existence of these hypothesized protagonists. Experiments to find the aliens are being conducted because serious, scholarly researchers think that they might succeed. They are convinced of the aliens' existence—not by the fantastic arguments of those who believe that UFOs come from the stars, but by a very ordinary argument.

AN ORDINARY SITUATION

There is a point of view among astronomical researchers that is generally referred to as the Principle of Mediocrity. This is not a reference to the astronomers themselves, but rather to our place in the universe. The Principle insists that there's nothing remarkable, nothing the least bit special, about our cosmic situation.

This cosmological modesty is a relatively new thing. Until four centuries ago, most Western societies subscribed to the Aristotelian world model: the Earth was the pivot of the universe, ground zero for creation. Such a view is unabashedly self-centered, but also easily justified by the obvious fact that the heavens appear to revolve around our planet. However, upon closer examination, this daily revolution is found to be imperfect. In particular, the Sun, moon and planets wheel across the skies in changing, complicated ways. In 1543, this heavenly complexity motivated Polish astronomer Nicolaus Copernicus to carefully set down arguments for yielding cosmic priority to the Sun. The movements of the planets, he noted, could be more simply understood if it was assumed that they orbited the Sun, not the Earth.

Copernicus died the year his treatise appeared in print, a case of publish *and* perish. But an early demise at least spared him the ensuing storm of controversy. Copernicus had displaced the center of the universe 93 million miles, not much by current standards. Nonetheless, his ideas met resistance, primarily because of their philosophical implications. He had dared to question the central role of mankind. If the universe was constructed for our benefit (and everyone who was anyone in the 16th century believed it was), why didn't we have a downtown address?

Relocating the crux of the cosmos to the Sun, despite the controversy, proved only to be the first of many moves. A paradigm shift in cosmology was underway, and the Sun had little time to enjoy being the center of the universe. By the beginning of the 17th century, astronomers knew that the stars were also suns, only farther away. Attempts to determine just *how* far away were initially stymied by the unexpectedly vast distances of even the nearest of these luminous pinpoints. But it was dismayingly clear that the traditional picture, in which the stars were lights attached to a mammoth sphere (and therefore all at the same distance), was desperately wrong. The stars were sprinkled across enormous spaces, and extended without limit towards an unseen horizon. Our sun was no more special than any of the stars. We were a speck of foam on a huge ocean. The comfort of a small universe was gone, and Blaise Pascal wrote that he was scared by "the eternal silence of infinite spaces."

Little did Pascal realize that the intimidating cosmos he knew was only a small fragment of what it would become. Telescopes grew larger, and the perceived size of the universe grew with them. The farther we could see, the more there was *to* see. As the 20th century began, the Dutch astronomer J.C. Kapteyn attempted to put all this into perspective, to establish the gross features of the cosmos. He concluded that the universe consists of a huge, flattened disk of stars, about 25,000 light-years across: the Milky Way galaxy. Kapteyn's size for the Milky Way was an astounding twenty million times greater than the distance across our solar system. The planetary assemblage that had been perceived as the totality of creation from time immemorial—the bodies extending from the Sun past Earth, Mars, Jupiter, Saturn and out to lonely Pluto—all this was now no more than a pinhead somewhere in Paris. Well, actually not just somewhere. According to Kapteyn, the solar system was very near the Milky Way's center.

By the early 1920s, even this was proven wrong. Kapteyn had misjudged the dimensions of the Milky Way. It is four times the size he measured, home to nearly a half-trillion stars. The solar system turns out to be located closer to its edge than its center. Astronomers had once again humbled mankind. The breadbasket punch to our hubris was this: the Milky Way is only one of many galaxies. Very many. Judging by a recent Hubble Space Telescope result, approximately 50 billion galaxies litter the visible universe. Each of these galaxies has its own complement of hundreds of billions of stars, on average.

It boils down to this. For four centuries, one astronomical discovery after another has moved us farther from the cosmic center stage. Our planet is but a small ball of rock around an ordinary star, one of ten thousand billion billion billion stars in the universe. Our astronomical situation is very ordinary, very mediocre. In comparison to the pre-Copernican situation, we have been seriously cut down to size. We have eaten a big slice of humble pie.

The popular reaction to this cosmic down-sizing has been defensive. In an attempt to maintain a bit of dignity, many folks rationalize our unremarkable physical circumstances by claiming that we are biologically, intellectually, or morally special. Our astronomical situation is ordinary, but humans are extraordinary. This thought underpins most of the popular reports of invading extraterrestrials. Sure, our planet may only be a dust mote in a raging storm, but we're still where the cultural action is. The aliens will want to come here (and maybe even have), if only to blast our cities with their ray guns.

Such a self-centered point of view is understandable, despite being an obvious attempt to resurrect those days when Earth was creation's nub and hub. But even though it may be personally satisfying to say that we are biologically or culturally special, it's unclear what that gets us. Certainly not cosmic neighbors. Sure, we may find a reason why aliens would *want* to interfere with our lives and our wheat crops, but few scientists believe that they have. The simple fact that cultured creatures exist on Earth doesn't seem to provide any reasonable basis for expecting that alien societies dot heaven's starry vault.

But with a slight attitude adjustment, it might. If we insist on being special, and if we allow that the aliens are *also* special, then we could invoke an old argument for why the universe might be widely populated. In ancient times, it was frequently stated that distant worlds (such as the Moon and Mars) were peopled not because life had sprung up in suitable environments there, but because these orbs had been deliberately fashioned for their inhabitants. The Creator had life on His mind, life was His grand design. Living bodies were His protagonists, and cosmic bodies played a supporting role. Creatures came first, and planets second. So the gleaming lights of the night sky must be inhabited. After all, what purpose would the Moon and Mars serve if they were unpopulated?

The modern counterpart to this "economy of worlds" argument is to note that our lives would be very little affected if some far-off galaxy were to be suddenly obliterated. By itself, a distant galaxy has only minuscule consequences for human existence. So why, then, has God bothered with the construction of that galaxy, with its hundreds of billions of stars and countless planets, if not for the benefit of *its* inhabitants? To let such a mammoth hunk of cosmic territory remain sterile would be "a waste of space," as Ellie Arroway frequently intoned in the movie *Contact*.

For two thousand years, this line of reasoning has been used to postulate a universe teeming with life. The cosmos is too big a house for humans to occupy alone. However, most contemporary scientists are reluctant to assume that God functions with the logic of a real estate developer, that He operates on the premise that a galaxy of worlds is useless if not inhabited. So those scientists who are seriously seeking the extraterrestrials justify their search with a somewhat different argument. It goes like this: we are astronomically mediocre. The makeup of the universe is the same from place to place. Light from the most distant objects visible to our telescopes confirms that the physics and chemistry of the universe's remotest realms is the same physics and chemistry we observe here. If, from the point of view of the physical sciences, we are ordinary, then perhaps we are biologically ordinary too. If the Sun and its retinue of worlds is only one system among many, then many other systems will be like ours: home to life.

Indeed, to the extent that this is true, we should be prepared for the possibility that, even in the Milky Way galaxy, billions of planets may be carpeted by the dirty, nasty business known as life.

For most scientists, the Principle of Mediocrity argues for abundant biology. Living creatures will fill the universe not because we are special, and not because *they* are special—but simply because the processes that give rise to life are *not* special, and therefore operate on untold worlds.

FINDING INTELLIGENCE

The cosmos could be cluttered with living creatures. But the greater question, the question of this book, is: do we share the universe with *intelligent* life? After all, when Captain Kirk boldly goes where no humanoid has gone before, it's not to search out new strains of bacteria or a better moss. He's looking for life that can hold up its end of the conversation.

Do such intelligent beings exist? Obviously, on Earth they do. So a further application of the Principle of Mediocrity should ensure plenty of clever cosmic company. Smart creatures walk the Earth, so they must exist on other worlds as well. However, mediocrity is a less convincing argument when applied to savvy creatures than when applied to biology in general. Life has flourished on this world for at least 80% of the time since our planet was born. But intelligent life has tramped Earth's surface for only 0.02% of that history. Life has had a long run, but humans are new to the stage. More to the point, the show might have played without their appearance at all. If the asteroid that wiped out the dinosaurs and most other species 65 million years ago had been a half-hour later getting here, it would have missed our planet altogether. In that event, mammals would have remained bit players on life's stage. Earth would be dominated today by large, boorish reptiles.

In other words, it's possible that life could be common, but that clever life might be rare. Invoking the Principle of Mediocrity suggests to us that the Milky Way supports a great deal of metabolism, but says little about E.T.'s IQ.

However, even if it turns out to be true that the evolution of intelligent life is uncommon, even if the majority of life is dumb, there is still abundant incentive to follow Captain Kirk's lead and search for our brainy brethren. Intelligent beings can create artificial habitats, such as spaceships, that enable them to move beyond the limited confines of their native planet. Extraterrestrials that are capable of interstellar travel could spread widely, even if intelligence was germinated on a relatively limited number of worlds. The Galaxy might be replete with aliens that long ago left home. In addition, modern science has given us techniques for detecting intelligent beings at great distances. We don't require that they leave their home turf to visit ours. If they signal across space with light or radio waves, for example, we have the technology to listen in. This is an intriguing possibility, for it doesn't require that either we or the aliens make long, expensive trips.

Today, a small group of scientists is attempting to eavesdrop on alien communications, searching for the subtle evidence of intelligent life elsewhere. So far, they've found nothing. Nothing. But the endeavor is a yes-or-no, either-or experiment. The search is a failure until that moment when it suddenly becomes a success. Many scientists, reputable scientists, believe that we might be in communication with intelligent beings from other star systems within our lifetime.

This is the first time such a possibility has existed. One hundred thousand generations of humans have walked the planet before us, a sea of ancestors who looked at the night sky and wondered if somewhere in those darkened spaces there were other thinking beings looking back. It is overwhelmingly likely that as long as man has existed, he has shared the universe. But ours may be the first generation to actually uncover evidence for this fact, and find the unmistakable traces of sentient neighbors.

In the last four centuries, astronomers have compelled us to accept that we occupy a humble, unremarkable place in a vast cosmos. No longer the center of the universe in an astronomical sense, we continue to take refuge in our earthly intellectual and moral primacy. The discovery of intelligent signals from space would remove these final vestiges of pre-Copernican insolence. We would know that we are neither culturally nor intellectually supreme, not the crown of creation, but simply one self-aware, thinking society among many. We would no longer be special. Nor would we be alone.

No one knows if this will happen, or when. But it could be tomorrow.

Aliens in the Backyard?

If we share the universe, how close might the neighbors be? Throughout history, recorded and otherwise, most speculation about cosmic companions has centered on those few heavenly bodies that attract our attention by their size or movement. The planets, for example, stand out as bright stars that wander through the skies. The moon and Sun dominate the heavens. This handful of conspicuous objects also happens to include Earth's nearest cosmic neighbors. If we believe that the skies harbor alien beings, it is only logical to begin our search on these local worlds.

CRATER CREATURES

Not surprisingly, the first other astronomical body examined for tenants was the moon. As early as the first century, the Greek essayist Plutarch remarked on the moon's dark and light spots, features visible to the naked eye. He assumed that they were land and oceans. Plutarch even claimed to see mountains, which bespeaks either exceptional vision or substantial imagination. He concluded that the moon was rather like the Earth, and populated as well.

During the Middle Ages, Plutarch's optimistic ideas became politically incorrect. Lunar inhabitants were ruled out on theological grounds. A strange, self-centered twist was made on the classical argument that worlds were constructed for the benefit of their occupants. The new logic ran like this: Since the universe had been constructed for mankind's use, to populate the moon would be to befoul its purpose as a celestial ornament for our enjoyment and wonder. Humans were too important to share the universe. Even the lunar dark and light spots seen by Plutarch (and anyone else who cared to look) were dismissed as defects of vision. Medieval religion proclaimed the moon to be a perfect and unblemished sphere.

As the Renaissance dawned, the moon reverted to its former, imperfect state. Galileo turned his crude telescope skyward, and saw a rugged lunar surface peppered with giant potholes, high terrain, and gray, sea-like patches. The blemishes, better known as topography, were unmistakable. In 1610, the same year that Galileo published his observations, the German astronomer Johannes Kepler, who had also peered moonward with a telescope, made a startling pronouncement regarding a large, circular lunar feature. It was, he claimed, a construction of the local inhabitants, who were also its occupants. (In fact, it was probably one of the larger craters.) Having been among the first to behold the intriguing landscape of our natural satellite, Kepler found it only a small step for a man to speculate that it was rife with life.

This speculation was, of course, naive. Nonetheless, until quite recently confidence abounded that aliens, or at least life, ran rampant in our immediate backyard. The enthusiasm for neighborhood extraterrestrials flourished in Victorian times. During the 19th century, telescope technology substantially improved, giving astronomers far sharper views of the moon and neighboring planets. Some of these sibling worlds, only poorly glimpsed before, now showed intriguing and occasionally suggestive detail. A new literary genre, later to be called science fiction, titillated the public not merely with the pronouncement that these objects were inhabited, but with descriptions of the occupants' behavior and appearance. There was more than just life in the solar system; there was intelligent, technically sophisticated life. The writers who capitalized on the new narrative form quickly realized that the interesting thing about extraterrestrials was not so much their at-home lifestyles as their interaction with us. So sci-fi stories (and in this century, sci-fi films) relied heavily on interplanetary travel to bring curious humans to these other worlds or, if the author had a more apocalyptic bent, to bring the aliens to us.

In 1865 a young Frenchman, Jules Verne, penned *From the Earth to the Moon*, a story often promoted as science fiction's pioneering effort. Verne's travelers were shot lunar-ward by a patently absurd, super-sized cannon (situated in Florida, in an unintended foreshadowing of Cape Canaveral). With no control of their ballistic spacecraft, these Victorian astronauts couldn't actually land on our cratered neighbor, but merely enjoyed a quick, circumlunar slingshot tour. Verne thereby spared himself the inconvenience of having to describe the moon's inhabitants, if there were any.

A Victorian contemporary, H. G. Wells, was less reluctant. In his 1901 novel, *The First Men in the Moon*, Wells levitates his travelers to their extraterrestrial destination by use of a special mineral capable of canceling gravity. Once there, they meet up with a sophisticated lunar civilization composed of insect-like Selenites. Moon melodrama had become popular, and a year later, Georges Melies, a special effects pioneer, made the first science-fiction film, a 16-minute effort entitled *Le Voyage de la Lune.* His story incorporated both the big-barreled blastoff of Verne, and moon dwellers *á la* Wells. Melies' crater creatures were approximately humanoid, sporting chicken heads and lobster hands.

Of course, that was all fiction. On the moon, the situation was less dramatic and the action minimal. For three billion years, only the noiseless impact of an occasional meteor had punctuated the moon's dull existence. Here was an inert world, lacking atmosphere and sound, in slavish orbit around a master that was its complete opposite —a watery planet where change abounded, and where life flourished and evolved. But in 1959, the moon's monotonous solitude ended. A crude Soviet spacecraft crashed silently and unseen onto the stone-gray surface, the first man-made object to make the trip. A decade later, on 20 July, 1969, a spidery assemblage of painted metal appeared in the black sky above the Sea of Tranquillity. It dropped toward the moon on a mute tail of flame. Moments later, the awkward-looking device kicked up some ancient dust and settled onto the surface. Mankind had finally come to the moon.

Neil Armstrong climbed slowly out of the lunar lander, looked away from the harsh sun, and took a small step for a humanoid and a big step for mankind. He was greeted by stony, dusty silence. The folks that sent Armstrong to the moon hadn't anticipated any welcoming party, no delegation of Selenites offering him the key to their crater. They expected, and the Apollo astronauts found, an airless, waterless world—a hundred million trillion tons of dead rock. Armstrong and fellow crew member Edwin Aldrin had brought their own atmosphere, their own food and drink. They, and the ten astronauts who would follow, shuffled about the dusty lunar surface utterly alone, plucking rocks from the lifeless landscape. They didn't concern themselves about chicken-headed locals that might be hidden in the moon's pitch-dark shadows. They didn't bother to bring jars for bottling specimens of the flora and fauna. None were anticipated.

Despite the fantasies of Victorian novelists, astronomers had known for more than a century that the moon had no atmosphere. It's an easy thing to know, in fact, and the skeptical reader can make a simple check. Due to its monthly orbit of the Earth, the moon slides across the sky relative to the stars, moving approximately its own diameter every hour. Occasionally, the moon will occult a distant star, an event that can be observed on a clear night with binoculars and a little patience. The moon approaches the star and then douses it abruptly. There is no gradual dimming while the hapless star gets closer to the moon's limb, such as would occur if the moon had any substantial atmosphere. In fact, the gas density above the moon's cratered surface is ten trillion times less than Earth's, an impressive vacuum.

The building blocks of life are also in short supply on this pockmarked world. There is little carbon or oxygen in the crust. The essential ingredient for biological activity, water, wasn't found by the Apollo astronauts, not even chemically present in the boulders brought back from the moon. Yes, radar measurements by the American Clementine space probe suggest that a small field of dirty ice lies near the lunar south pole (although the result is controversial). But even if it exists, this dusty, chunky frozen lake would be exceptional: the preserved remains of a crashed comet, fortuitously located in everlasting shadow at the bottom of the moon. Water is in excruciatingly short supply on Earth's partner world.

The lunar rocks have also been probed for life, including, of course, fossil life. None has been found. The moon is sterile, and so obviously devoid of biology that after the third Apollo landing, program directors stopped requiring the astronauts to sit in quarantine after returning to Earth. Despite the lunatic claims of Plutarch and the fanciful predictions of turn-of-the-century sci-fi writers, it appears that the only life that has ever existed on the moon was brought to it courtesy of NASA and was dressed in space suits.

SISTER PLANET

The moon boasts no crater creatures, no Moon Maid *á la* Chester Gould. Our nearest cosmic neighbor is as lifeless as a doorknob. But that's neither a surprise, nor much of a loss. There are, after all, eight other planetary worlds and several dozen moons in the solar system: habitats in abundance. So when people who sought aliens in the neighborhood became discouraged by the moon's rigid stone face, they refocused their telescopes and their attentions on nearby planets. The closest of these is Venus.

At first blush, the idea of Venusian life seems reasonable. Once dubbed Earth's sister planet, Venus is nearly identical in size to our own world. At closest approach, the sisters are a mere 26 million miles apart. You might think this proximity would allow good telescopic views of the Venusian topography. And it would, except for the fact that the weather report for Venus is cloudy. Today and every day.

Venus modestly hides her surface with a cloud cover that never clears and is never broken. It is seductive to imagine that under this secretive canopy lies a moist, swampy world, crowded with life. Science fiction has encouraged this view. Supermarket tabloids occasionally print "revealing NASA photos" showing dinosaurs and other implausible critters cavorting in the dull light of a Venusian day. In 1968, Hollywood sent *Zontar, the Thing from Venus* to cavort in third-rate movie theaters. A real Zontar, however, would have required an asbestos epidermis. Astronomers of the time correctly suspected that Venus' perpetual cloud cover had turned the planet into the solar system's largest hothouse.

Those suspicions were aroused when the Mariner 2 space probe made the first successful flyby of Venus in 1962. Mariner measured air temperatures of 750 F. In addition, Venus' atmosphere was found to be thick and poisonous, consisting almost entirely of carbon dioxide laced with sulfuric acid for extra bite. The heavy atmosphere crushes down on the planet's surface with nearly one hundred times the pressure of Earth's air.

In 1975, a Soviet probe, Venera 9, was parachuted into Venus' opaque clouds. As it fell towards the surface, the temperature and pressure of the white fog outside the craft climbed. Two dozen miles above the planet's hellish landscape, the clouds broke and the air suddenly cleared. Minutes went by, while the probe dropped through the searing atmosphere. Venera 9 came to rest in a rock field and took the first picture of the surface. The local temperature was 900 F. Within an hour, the probe died of heat exhaustion.

Three days later, another Soviet lander managed to snap a photo of Venusian real estate before it, too, was crushed and cooked. These two photos remain among the small handful we have of Venus taken from the ground. Neither dinosaurs nor Zontar are visible. Venus is an oven—indeed, it is hotter than an oven. The choking, corrosive atmosphere is so effective at blanketing the planet that even during the two month-long Venusian night, the landscape remains faintly illuminated by the eerie, red glow of hot rock. Venus may be Earth's sister planet, but it's not the sister you'd like to spend more time with.

There may have been a time, however, when Venus was a nicer place to call home. Four billion years ago, when the solar system was still in its infancy, the Venusian climate was possibly more temperate, if for no other reason than that the juvenile Sun was only about two-thirds its present luminosity. And there's reason to believe that Venus was once blessed with as much water as Earth. In earliest times, all the inner planets were hot and semi-soft, still molten from the trauma of birth. For a few hundred million years thereafter, they were stuck in a dangerous, hellish place, ducks in a giant shooting gallery. Asteroids and comets, the left-over pieces from the planets' creation, careened wildly through space, and smashups were common. Indeed, it was these collisions that ultimately cleaned out the solar system, and killer comets, once a constant threat, eventually became a rare one. But while this rain of rock and ice temporarily interfered with the development of life on Earth and possibly elsewhere, the resulting debris provided these worlds with atmospheric gases and water. The oceans of Earth are filled with the juice of ten thousand crashed comets the size of Hale-Bopp. A similar liquid endowment was bequeathed to Venus. Our sister planet once had oceans.

Could life have once sprung up in these long-gone Venusian seas before they, and it, were boiled away by the planet's runaway greenhouse effect and the Sun's growing brilliance?

It was this possibility that contributed to the excitement attendant upon the launch of NASA's Magellan spacecraft in 1989. Thrown into orbit around Venus, there was hope that Magellan might uncover evidence of Venus' fecund youth, if it had one. The craft used its penetrating radar to slowly build up a picture of the planet, defrocked of its obscuring clouds. It revealed a strange topography that included mountainous highlands, puzzling creases, "pancake" patterns, and large fields of volcanoes. Disappointingly, there were no dry river beds or empty ocean basins offering testimony to a time when water had flowed and splashed on Venus. Such tantalizing evidence of a wet past, if it exists, may be irretrievably lost. Active volcanism on Venus has steadily resurfaced the planet, burying whatever tales her face may have told of a distant and more hospitable history. Venus is determined to hold on to her secrets.

Yes, Venus was born Earth's twin. But she came into existence one-third closer to the Sun than our own planet. For this inside track, she has paid an awful price. Her primeval oceans were warm—so warm, in fact, that like seltzer water left on the counter top they soon went flat. Their natural carbonation, the dissolved carbon dioxide (CO_2), was bubbled out. An efficient greenhouse gas, the CO_2 soon made Venus even warmer, and this caused yet more of the gas to gurgle out of the tepid seas. Eventually, conditions on Venus got so toasty that the oceans simply boiled away, leaving Venus today a sterile desert, roasting silently under a suffocating, perpetually cloudy sky. There are no Venusians now. There probably never were.

A computer-enhanced image of a volcano (Maat Mons) on the planet Venus. Lava flows extend hundreds of miles across the fractured plains in the foreground. [NASA]

ANGRY RED PLANET

The moon and Venus aren't attractive for life, and few of the other planets tempt biology either. Tiny Mercury is hot and airless. Its searing temperatures are the inevitable consequence of being even closer to the Sun than Venus. With a gravitational tug only 38% that of Earth, its original atmosphere vanished long ago. The giant outer planets—Jupiter, Saturn, Uranus and Neptune—are cold, gassy balls of methane and ammonia. In this array of generally forbidding worlds, only one offers obvious enticement: Mars.

Until the mid-18th century, Mars held no special appeal either for astronomers or the public. Indeed, it was less intriguing than the moon, with its pockmarked face, or Saturn, with its dramatic rings. Then along came a highly skilled champion for the red planet: William Herschel. Early in his career, Herschel had a day job playing the organ at the Octagon Chapel in Bath, England, and studied the stars on evenings and weekends. He later turned professional, accepting a position as astronomer for George III. Herschel was arguably the greatest observer of all time. In addition to a crammed schedule that included building large telescopes, cataloging stars and nebulae, and discovering Uranus, Herschel spent many a night viewing Mars. He studied its polar ice caps (which he decided were snow), and established that the red planet had a day very similar in length to our own: 24 hours and 37 minutes. In a talk given to Britain's Royal Society in 1784, Herschel claimed that Mars possessed an atmosphere, and consequently "its inhabitants probably enjoy a situation in many respects similar to our own." With the mention of "inhabitants," the possibility of living beings on the red planet had been broached by a respected scientist in an academic setting. Martians were assumed to exist.[1]

During the following century, Mars rose to the top of the astronomical popularity charts. After all, it was the only planet on which surface detail could be readily seen. Herschel had spoken of permanent spots on the planet, and in the 1860's Angelo Secchi in Rome noted some peculiar, streaky features as well. But it was the work of Giovanni Schiaparelli, director of the Milan Observatory, that transformed the perception of Mars for a hundred years. In 1877, Schiaparelli claimed to see a network of 79 linear features which he called *canali*—translatable as "channels," or more commonly (and more provocatively), as "canals."

Were these merely natural formations, or the deliberate work of commercially-oriented beings? Schiaparelli himself thought both explanations were possible. His controversial claim was publicized widely, and in 1893, Percival Lowell, the scion of a prominent Boston family, became infected with canal fever. Lowell decided to abandon his career with the family business and turn his talents to astronomy. He would soon be the canals' greatest proponent.

Lowell was no fool. He had a degree in mathematics from Harvard, and made many significant astronomical observations. But he is primarily remembered for his fascination with the canals. Lowell was aware that not all astronomers could make out Schiaparelli's *canali,*

---

1. It should be pointed out that Herschel was not overly particular when it came to populating the cosmos. As far as he was concerned, everything in the solar system was inhabited, including the moon and Sun.

but he attributed their regrettable inability to discern these constructs to bad "seeing," a term used to describe the clarity and stability of the air above an observatory. Seeing varies from place to place and from night to night. It is seeing that determines how much detail a human observer can glean through a telescope. If Schiaparelli could make out canals while others could not, perhaps the reason was that the air above Milan was less turbulent.

Lowell decided that the best seeing in the United States was to be found in Arizona. He resolved to set up telescopes there for the study of Mars, and did, situating his modestly-named Lowell Observatory in Flagstaff. For more than two decades, Lowell spent his nights behind the eyepiece, charting the trench-like handiwork of his putative Martians.

Lowell mapped nearly 200 canals onto small white globes, giving them Latin names for handy identification. His interpretation of the reason for all this civil engineering was straightforward. The canals were the work of an intelligent species, far more advanced than our own. The Martians dug their canals not for commerce, but for survival. According to Lowell, evaporated water from Mars' oceans had slowly escaped into space in consequence of the planet's feeble gravity. With their seas gone, the desperate Martians dug canals to bring vital water from the polar caps to their desert farms. Of course, narrow canals would be hard to see a few tens of millions of miles away, but the blue-blooded Bostonian theorized that the vegetation that grew along their banks made them appear wider and more conspicuous in Earth-bound telescopes (much as the Nile can be easily seen from space). It all made sense, and Lowell's story was enthusiastically received by the public, a circumstance that was encouraged by his considerable writing talents. Lowell's popular book *Mars,* which appeared in 1895, elaborated his case for a highly advanced society in the neighborhood. "Certainly what we see hints at the existence of beings who are in advance of, not behind us, in the journey of life," he wrote. "Startling as the outcome of these observations may appear at first, in truth there is nothing startling about it whatever."

As the twentieth century dawned, Mars—named for the Roman god of war—was widely thought to be inhabited by intelligent, accomplished, and clearly desperate beings. The consequences were manifold and, at least in the short term, menacing. Fictional stories of Martians appeared in profusion. In 1898, H. G. Wells' published the most famous of these, his cautionary tale, *The War of the Worlds.* In this novel, the antagonists are Martians that have graduated from trench work to planetary invasion. They come to Earth inside large, protective machines and proceed to lay waste major metropolises. The ruthless invaders are unfazed by our instruments of war, but are stopped by lowly bacteria, against which they have no natural immunity.

In the last hundred years, Wells' tale has been twice reprised, and with considerable impact. On Halloween, 1938, CBS's Mercury Theater broadcast a radio dramatization of *The War of the Worlds* under the direction of Orson Welles. In this version, the Martians make their malevolent appearance near Princeton, New Jersey (in the original story, London was the designated target), but by the end of the hour most of the U.S. is in their alien grasp. Despite a couple of disclaimers, approximately one million listeners to the radio play believed that the Earth was actually under attack by aggressive extraterrestrials. Part of this impressive gullibility can be attributed to news of very real invasions taking place in Europe as World War II erupted. Much of the rest was the durable legacy of Schiaparelli and Lowell.

In 1953, George Pal, the Hungarian puppeteer-turned-film-producer, made a highly-regarded movie version of *The War of the Worlds,* in which the Martians have set their destructive sights yet farther west, flattening Los Angeles before taking on the rest of the globe. This was mid-century. Wernher von Braun was designing rockets, the 200-inch telescope on Mount Palomar had been operational for years, and astronomers were convinced that the Martian atmosphere was asphyxiatingly thin. Nonetheless, the idea of intelligent aliens from the red planet—an idea that dated from Herschel's 18th century musings—was still credible to lay people.

A dozen years later, it wasn't. In 1965, NASA's Mariner 4 spacecraft flew to within 6,000 miles of the Martian surface, making a few dozen television-quality images of the landscape below. When these photos

The New York Times

Copyright, 1938, by The New York Times Company.

NEW YORK, MONDAY, OCTOBER 31, 1938.

PAT
EALER
ENATE

e Opposes
anges in
y Laws

OF TVA

nced, but
"Misery,"
imes

ve Mead's
ge 6.

pondent
et. 30.—Rep-
Mead, Demo-
e short-term
lection Nov.
a statement
campaign
THE NEW
r New York
ajor parties

n the main,
qualified de-
legislation
ber of the
ves, had a
passing.
Social Se-
or Relations

Radio Listeners in Panic, Taking War Drama as Fact

Many Flee Homes to Escape 'Gas Raid From Mars'—Phone Calls Swamp Police at Broadcast of Wells Fantasy

A wave of mass hysteria seized thousands of radio listeners throughout the nation between 8:15 and 9:30 o'clock last night when a broadcast of a dramatization of H. G. Wells's fantasy, "The War of the Worlds," led thousands to believe that an interplanetary conflict had started with invading Martians spreading wide death and destruction in New Jersey and New York.

The broadcast, which disrupted households, interrupted religious services, created traffic jams and clogged communications systems, was made by Orson Welles, who as the radio character, "The Shadow," used to give "the creeps" to countless child listeners. This time at least a score of adults required medical treatment for shock and hysteria.

In Newark, in a single block at Heddon Terrace and Hawthorne Avenue, more than twenty families rushed out of their houses with wet handkerchiefs and towels over their faces to flee from what they believed was to be a gas raid. Some began moving household furniture.

Throughout New York families left their homes, some to flee to near-by parks. Thousands of persons called the police, newspapers and radio stations here and in other cities of the United States and Canada seeking advice on protective measures against the raids.

The program was produced by Mr. Welles and the Mercury Theatre on the Air over station WABC and the Columbia Broadcasting System's coast-to-coast network, from 8 to 9 o'clock.

The radio play, as presented, was to simulate a regular radio program with a "break-in" for the material of the play. The radio listeners, apparently, missed or did not listen to the introduction, which was: "The Columbia Broadcasting System and its affiliated stations present Orson Welles and the Mercury Theatre on the Air in 'The War of the Worlds' by H. G. Wells."

They also failed to associate the program with the newspaper listing of the program, announced as "Today: 8:00-9:00—Play: H. G. Wells's 'War of the Worlds'—WABC." They ignored three additional announcements made during the broadcast emphasizing its fictional nature

Mr. Welles opened the program with a description of the series of

Continued on Page Four

OUS
RE
AR

Exile
or

REV

Othe
Pen

w
WA
evacu
thous
cordi
12,000
the
porte
after
front
borde
terri
Pol
offic
Com
relat
spec
tains
less
tion
cars
open

were given to President Lyndon Johnson, he looked at them with surprise. Mars' surface was littered with craters, not canals. The red planet appeared as lifeless as the moon.

Percival Lowell had been wrong, horrendously wrong. His eye had done what all eyes are good at doing: connecting the dots. Random small features glimpsed indistinctly through the telescope (in which Mars appeared no larger than a marble seen from across a room) became connected in Lowell's brain. That was the only place where the canals existed.

Suddenly, intelligent Martians, who had populated books, movies, and radio plays for nearly a century, were no longer thought to populate Mars. Life on the red planet was still possible, of course, but scientists and most of the public demoted it to something a little less dramatic than Wells' invaders. Martians, if they existed, were thought to be of a simpler form: primitive plants and animals that could withstand the rigors of a world where temperatures were cold, and liquid water was absent.

Since low-grade Martians weren't likely to invade Los Angeles or be visible in our telescopes, the only way to find them was go to the planet's surface. On July 20, 1976, precisely 7 years after Neil Armstrong's first lunar foray and nearly a century since Schiaparelli's description of *canali,* the thin skies above Mars were broken by a NASA space probe. The Viking lander touched down without difficulty on the Chryse Planitia, a sprawling, rock-strewn plain 800 miles north of the equator. It was followed two months later by its twin, Viking 2, which landed on similar terrain nearly halfway 'round the planet.

When the Viking landers' cameras opened their eyes in the frigid Martian air, no one was sure what they would see. Would there be strange plants or exotic, three-eyed creatures with green complexions arrayed in front of the lens? The tracks of animals or insects? Possibly even the ruined remnants of an earlier civilization? All of this was

Front page story from the New York Times regarding Orson Welles' 1938 radio broadcast of *War of the Worlds* and the chaos it caused. [New York Times]

possible. But what the cameras saw was far starker: red fields of rocks stretched out against a pinkish sky. Small boulders and dusty sand. There were no creatures staring back at the cameras, and no plants huddled in the weak sun. For months the photos continued to come in, but the view scarcely changed. No animal tracks appeared. Nothing grew, other than some occasional patches of morning frost, and nothing moved, other than the wind-blown dust.

This photo of the surface of Mars was taken by Viking Lander 2 in 1979, and shows a thin coating of water ice on the rocks and soil. [NASA]

Undaunted, the Viking landers pressed ahead with other attempts to find life, efforts that were subtler than simple visual reconnaissance. Robot arms scooped up samples of the surrounding dirt, even pushing aside rocks to get at shaded soil that was less likely to be sterilized by the brutal ultraviolet rays stinging the surface. The samples were brought into the spacecraft and given a battery of tests designed to detect any microbial life within. Nutrient liquids were poured on the Martian dirt, and sensitive instruments looked for exhaled gasses that would signal metabolism. They found them. The nutrient-soaked soil was expelling large amounts of oxygen and carbon dioxide, and the landers telemetered the news back to eager scientists on Earth.

Needless to say, this occasioned considerable excitement. But the celebratory champagne flowed only briefly. It was soon suspected that the landers had detected chemistry, not biology. Some of the critical tests for life had come up empty. During the metabolic testing, mass spectrometers aboard the craft had scrutinized the soil for organic molecules, the basis of all life on Earth. They found none, to levels of a few parts per billion. On our planet, even the nearly lifeless soils of Antarctica would have yielded abundant quantities of these biological building blocks. Some of the Martian soil samples were baked for hours at 350 F, and tested again. The reactions persisted, despite temperatures that would surely have cooked any mini-microbes.

What was the bottom line? The rate of the reactions, their strength, their persistence after baking, and the failure to find organic molecules convinced nearly all the Viking scientists that they had failed to find life. The Martian surface is dry, and scorched every day by the ultraviolet-rich sunlight that penetrates the planet's thin atmosphere. This leaves it in a highly-charged, non-equilibrium condition. When you add water (in the form of nutrient solutions) to this deprived soil, some fast chemistry results.

Did this mean, as many reluctantly concluded, that Mars is biologically dead? Probably, but not certainly. After all, the Viking landers were only a quick reconnaissance for life, at least partially motivated by the fear that the Soviets would get to the red planet first, and (literally) scoop America on the scientific find of the century. With only two landers, immobile after touch-down, the amount of Martian soil subject to testing was small. Additionally, the landing sites were chosen for reasons that had little to do with life. It was more important that the touch-downs be on relatively smooth terrain, so that the landers would not catch on a boulder at the last moment, tip over, and disappoint millions of American taxpayers.

There were even some who argued that Viking's biochemical tests were misguided to begin with. They were designed to detect only "life as we know it." Maybe the Martians had some other, less obvious biochemistry. Maybe. But few project scientists subscribed to such a theory. They might just as well believe that the Martians were, in fact, clearly visible in the photos, but cleverly camouflaged as rocks.

It is a peculiar irony of the Viking missions that while the landers were vainly poking around the dead, dry sands of the surface, the orbiters that had dropped them there were making aerial photos suggesting that life may once have glutted Mars. Tirelessly circling the planet, the orbiters snapped thousands of photos of the red planet, the best we had ever had—indeed the best we *would* have for more than twenty years.

These photos showed a surface littered with craters, mountains, and immense, dead volcanoes. But they showed something else: systems of gullies and innumerable channels snaking over the landscape like dry rivers in the American west. These looked like the desiccated footprints of water that long ago ebbed and flowed across the landscape. Some of the serpentine features are clearly the result of sapping, the collapse of the surface into tunnels carved by the action of underground water. Others might be due to the sudden welling up of water from buried aquifers heated into action by volcanism. The largest surface cleft, Valles Marineris, often described as the Grand Canyon of Mars, is a rift caused by tectonic forces.

But such special circumstances aside, most of the photo evidence pointed one way. The bulk of the canyons and ravines are the remains of ancient branched river systems, where rain was collected and channeled. The orbiter photos provide compelling testimony to the fact that Mars was once a wet planet, on whose surface water coursed freely, and oceans and lakes pooled and stood.

That happy time was in the distant past, 3.5 to 4 billion years ago, when the planet was still young. Possessed of a thicker atmosphere and warmer temperatures, early Mars resembled early Earth. The resemblance lasted for hundreds of millions of years. But then things began to sour. As on Venus, some of its atmospheric carbon dioxide inevitably dissolved in the oceans. The acidic character of this Martian seltzer water was capable of weathering rocks. The gas then combined with the weathered rock to produce carbonates, which slowly accumulated as limestone sediment on the ocean floors. Unlike on Venus, the greenhouse gas was slowly *taken* from Mars' atmosphere, and turned into useless wallboard at the bottom of its lakes and oceans.

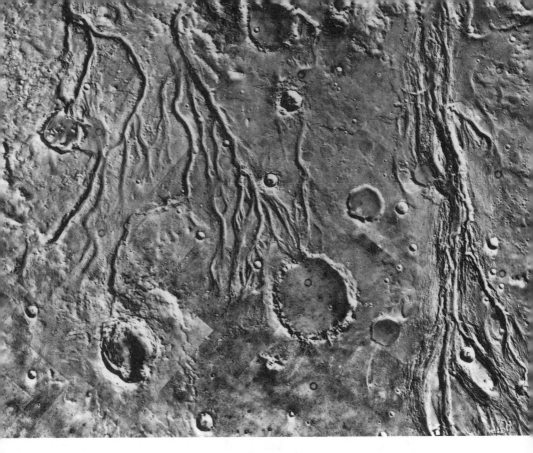

These channels on Mars were photographed by Viking Orbiter 1 in 1980. They suggest that a large volume of water once flowed on the Martian surface during a warmer and wetter epoch in Mars' distant past. [NASA]

On Earth, this process is compensated by the motion of continents. As land masses slide over the ocean floor, they stir and melt its sedimentary layers. Volcanoes then spew much of the melted material, including the carbon dioxide, back into the atmosphere. But Mars is a small planet, only half the diameter of Earth. It quickly lost the heat from its interior, energy left over from its incendiary birth. The fires from its volcanoes were banked long ago, and its continents remain rigidly fixed in place. Recycling of the wallboard-like residue of its greenhouse gasses stopped. Despite a promising start, the red planet freeze-dried. The temperatures dropped, the liquid water so essential to life froze hard in the permafrost, and the atmosphere became thin and wispy. Venus turned ugly because it was a little too close to the Sun. Mars went bad because it was a little too small. This is not the scenario that Percival Lowell described when he spoke of the death of the red planet, but his tales of a dying, drying world were remarkably prescient.

Despite the unattractiveness of present-day Mars, the Viking orbiter photos had provided evidence of a halcyon period in its history—perhaps a half-billion years, during which conditions on the red planet and Earth were comparable, and before the Martian freeze set in. In those long-ago times, single-celled life developed on Earth. Could not the same have occurred on Mars? Could it be that the red planet was not always a dead planet?

Possibly. In the summer of 1996, scientists from NASA and Stanford University made the astounding claim that they had found evidence of life from the fabled planet's youth. The evidence was retrieved not from Mars, but on Earth. The researchers broke open and analyzed a small meteorite picked up in Antarctica and thought to be flung from the Martian surface long ago. Gases trapped in the meteorite matched the composition of the Martian atmosphere as measured by the Viking landers. There is little doubt that the Antarctic rock is from the red planet. It is a piece of ancient, Martian crust, nearly four billion years old. Within the meteorite, which is about the size and shape of a large potato, the scientists found chemical compounds that are known to be produced by simple life. They also found what looked like microfossils —tiny structures reminiscent of bacteria. Small swarms of bacillus-like shapes, frozen in the rock, were also revealed by the microscope. Another photo showed a tubular form resembling a miniature, single-celled worm. Are these creatures from a primeval time, 3.5 billion years ago, when the solar system was young, and Mars was warm and wet? Have we found the former inhabitants of the red planet?

No one yet knows, and many scientists are skeptical. J. William Schopf, a paleobiologist at UCLA, was asked by NASA to comment on the meteorite research. He stated that it was "Interesting. Exciting. But circumstantial, inconclusive." Schopf pointed out that the evidence offered by the group claiming fossil life could be caused by natural processes, not biology. He would be more convinced if someone could find cell walls (admittedly a very difficult thing to see, even if they exist), or perhaps a few of the petrified critters embarrassingly caught in the act of cell division. In Cleveland, Ohio, a second research team that analyzed the same Martian meteorite concluded that the intriguing compounds found by the NASA and Stanford researchers were the result of natural chemistry involving the red planet's atmosphere, not the bodily processes of Lilliputian Martians.

Settling the matter of ancient life on Mars will certainly encourage the study of more meteorites, but the question may hang fire until we can examine Mars itself. While plans for rocketing human visitors to the red planet have been stalled for years, pint-sized robot rovers are lined up at the launch pad. These ambulating robots will scrutinize the

LEFT: This meteorite, labeled ALH 84001, was dislodged from Mars by a huge impact about 16 million years ago. It landed in Antarctica 13,000 years ago and was discovered in 1984. In 1996 scientists announced that they had found possible evidence of primitive life forms inside the rock. [NASA]

RIGHT: These electron microscope images of Martian meteorite ALH 84001 show the tube-like structures that may be fossils of primitive, bacteria-like organisms that lived on Mars more than 3.5 billion years ago. [NASA]

Martian landscape with infrared eyes and even small microscopes, in search of ancient inhabitants.

Ironically, the chances that these mechanical researchers will find any microfossils are enhanced by the death of Mars. The curse that turned the red planet into an arid icebox might be a blessing when it comes to looking for life's remains. The lack of tectonic activity ultimately cost Mars an atmosphere and its relatively warm weather. But it also ensured that the planet's surface remained intact. On Earth, the incessant movement of continents and the creation and destruction of the ocean floors has chewed up and spit out most of the rocks. But the sands and soils of Mars are old, and largely undisturbed. The microbial record, if it exists, should still be there.

So researchers are poring over the spacecraft photographs of the Martian landscape, looking for the best places to send the rovers. They seek out regions that have had persistent water: lake basins, or even the calderas of Mars' dead volcanoes, which, at least on Earth, are often the site of prodigious thermal spring activity. Terraced formations along streams beg to be explored, as do shoreline features that look like bathtub rings. The Viking landers of the 1970's were poorly equipped to find the carcasses of ancient life, and they didn't. But the small robots that will soon be doing a slow-speed crawl over the planet's icy landscape might.

The possible existence of archaic biology on Mars gives rise to an intriguing thought. Could it happen that the breadbox-sized rovers will uncover life that's neither old nor dead? Could they find *living* Martians, descendants of early life that, over the course of eons, have been able to adjust to the slow worsening of the weather on Mars' surface? Could there be microscopic creatures that have been able to retreat to a safe ecological niche, perhaps hundreds of feet underground, and survive the slow catastrophe that has taken hold of this world? In short, could there be contemporary Martians?

Despite the negative results from the Viking landers, this is not such a radical idea. After all, earthly bacteria have been found 3,000 feet underground, contentedly dining on water-soaked basalt and iron compounds, and ignorant of both sunshine and photosynthesis. On Earth, such buried life may be enormously abundant, possibly rivaling the quantity of microbial beings that live on the surface. When asked about the possibility that Martians are still thriving on the red planet, Jack Farmer, a NASA researcher involved with the search for minuscule Martians, offered an optimistic opinion. "Well, all I know is that on the Earth, bacteria have taken over just about every environment, including the subsurface ones," said Farmer. "They're everywhere, and they're very happy making a living without all the benefits of the surface. So if you ask if there could be Martians living today, I say yes, why not?"

The belief in Martians has flipped and flopped. At the beginning of this century, the red planet was thought to be occupied by intelligent beings able to irrigate their planet and possibly eradicate ours. At century's end, optimism about Martians has been renewed, although their assumed level of sophistication has been seriously downgraded. In H. G. Wells' story, the inspiration for which was the provocative work of Percival Lowell, the aggressive Martians were stopped in their destructive tracks when they fell victim to common bacteria. How ironic to think that we now believe that if the Martians exist, they *are* bacteria.

FAR-OUT LIFE

The possibility that Mars, despite its hostile surface environment, might still shelter primitive, underground life, has rekindled interest in other solar system habitats. Possibly our neighboring planets are not quite as unattractive as they first appear.

Consider Jupiter, by far the biggest planetary kid on the block. This outsized world has no solid surface, merely a swirling atmosphere thousands of miles deep covering a perpetually dark, hidden sea of liquid hydrogen. The heavy Jovian air is mostly ammonia and methane, compounds more suitable for heating your home or cleaning your bathroom, than fostering life. Sunlight striking the distant planet is 25 times weaker than on Earth, so temperatures at the cloud tops are unendurably frigid. Yet despite this long list of shortcomings, some scientists have seriously suggested that Jovian beings could exist. They point out that as one descends into Jupiter's cloudy atmosphere, the temperatures rise dramatically. At less than 100 miles below the cloud tops, comfortable room temperatures are reached. As astronomer Carl Sagan has wondered, why couldn't some sort of microbial life float in these temperate, misty layers?

Alas, this intriguing idea has been largely dashed by the results from NASA's Galileo probe, which parachuted into Jupiter's atmosphere in December 1995. The slowly descending probe sent back information on the composition of the Jovian clouds for nearly an hour before succumbing to heat and pressure. Its on-board mass spectrometer failed to find any complex organic molecules in the thick air. Additionally, the vertical wind motions were much higher than expected. Any floating life would be on a non-stop elevator ride between cloud layers that are unbearably cold and insufferably hot. So the idea that buoyant beings might live in the Jovian mists is now only improbable science fiction.

Still, there are other seductive sites for life nearby. The best of these is undoubtedly Europa, one of the four large moons of Jupiter discovered in 1610 by Galileo. Photographs by space probes show a smooth surface, crazed by thin, dark lines. Europa is fully covered in a sheet of ice. While interesting, Europa's frozen glaze is not surprising, given the frigid temperatures (-230 F) that exist so far from the Sun. This moon, presumably, has always had an icy crust. And yet, there are no craters marring Europa's glassy epidermis, no tell-tale pockmarks from the days when the solar system was a shooting gallery for asteroids and comets. This suggests that Europa's ice sheath has been renewed. Unlike our own moon, the face of Europa changes.

But what causes this satellite's dynamism? As it orbits its massive planetary master, Europa is subjected to repeated gravitational squeezing

Two views of Jupiter's ice-covered satellite, Europa, taken in September 1996 by the Galileo spacecraft. Long, dark lines are fractures in the frozen crust, some of which are nearly 2,000 miles long. [NASA]

and stretching by Jupiter and its sister moons. This constant abuse produces heat, much in the same way that kneading bread dough warms it slightly. For years, theoretical models concocted by planetary astronomers have suggested that Europa could be covered with a 60 mile-deep ocean, topped off with an icy crust. But the gravitational kneading would keep the water underneath above the freezing point, and liquid. Photographs by the Galileo spacecraft support this idea.

Might there be life in Europa's distant, dark seas? Biological beings that have evolved over the eons to accommodate themselves to the thin sunlight just under the ice, or perhaps to no sun at all? The somber idea of oceanic life without light has plenty of precedent on Earth. In the deep submarine trenches that mark the Pacific Ocean rift, hot-water vents spew life-sustaining warmth into the inky depths, sustaining colonies of bacteria. Long, snake-like tube worms feed on the bacteria. Perhaps similar undersea environments on Europa are also crammed with life. There are many researchers who believe that if you're looking for critters that are alive today, Europa might be a better hunting ground than Mars.

Some fictional aliens agree. In Arthur C. Clarke's story *2010,* sentient extraterrestrials from afar are well informed about Europa's seductive seas. The aliens gesture towards the planets and moons of our solar system, and magnanimously offer most of them for our inspection and use. Most, but not the lot. "All these worlds are yours," declare the aliens, "except Europa. Attempt no landings there." They repeat their message 93 times.

Europa may be off limits (although NASA is already planning to ignore the fictional aliens' warning), but at least these helpful beings have not erected a similar "No trespassing" sign on another beguiling moon, Titan. Saturn's largest satellite, Titan is comparable in size to the planet Mercury, and is the only moon known to have a substantial atmosphere. The view from Titan is spectacular, with Saturn and its rings arching across 20° of sky. The distant Sun is only a small, round dot. Actually, this is the dramatic view one would have if Titan's atmosphere were transparent, which it's not. The bitterly cold temperatures of this far-off world have produced a chilling, opaque smog that eternally blankets its surface. Water, of course, is frozen solid on Titan, as hard as granite. But astronomers theorize that ethane and methane, which would be liquid, might rain down on the moon's frigid surface, perhaps pooling into sticky lakes and oceans. If so, it's conceivable—albeit unlikely—that some sort of exotic life based on these low-temperature organic compounds might eke out an existence there. As part of NASA's Cassini mission to the ringed planet, a reconnaissance will be made of Titan. In 2004, upon the spacecraft's arrival at Saturn, Cassini will drop a probe to the sludge-covered surface of this enigmatic moon. During the three hour descent, the probe will sniff the atmosphere and check for the presence of interesting organic molecules. It is even expected that the probe will survive for a few minutes once it hits the viscous surface, and relay back photos that might establish the existence of any hydrocarbon habitants.

For thousands of years, mankind has optimistically speculated on the possibility of cosmic company nearby. It is a remarkable circumstance that those of us alive in the late twentieth century have finally had the opportunity to go beyond speculation, and actually explore many of the worlds of our solar system. We have walked the ancient dust of the moon, and made photos of Venus' scorching surface. We have sent our spacecraft to tantalizing Mars, for over a hundred years the postulated home of highly advanced beings, and our spacecraft eyes have shown us rock-strewn, empty deserts.

We have seen beautiful, terrible worlds, and no signs of sophisticated inhabitants. By the 1970s, two decades after the launch of Sputnik, few researchers continued to believe that other thinking creatures share our solar system. But they were hopeful that other life, perhaps only primitive life, was nearby. This modest expectation was dealt a body blow by the failure of the Viking landers to find any Martians, even small ones. For the next twenty years, a lot of researchers described the universe closest to Earth as D.O.A.

But reports of the solar system's lifelessness may be exaggerated. Now there is suggestive evidence that biology may have started its durable workings on Mars long ago, and at least two large moons of the outer solar system may sport conditions that, while unappealing to humans, could support life.

Although such thoughts are exciting to biologists, they are less comforting to those who wish for aliens that can hold up their side of the conversation. If there is life nearby, it consists of underground bacteria, simple creatures swimming in the pitch-dark oceans of a distant moon, or greasy, hydrocarbon slime. A half-century of space exploration has brought us to a conclusion that would have dismayed every previous generation: we have no *intelligent* company among our sister planets in the solar system. If we wish to find thinking beings, we must extend our search outward, to the vast stellar realms of the Galaxy.

Extending the Search to the Stars

The venerable notion that intelligent extraterrestrials might be lodged on other planets of the solar system is tough to give up. Close-by inhabited worlds have visceral appeal. They could also present us with some enticing opportunities. Establishing two-way communication with adjacent aliens would be technologically easy. So would interplanetary travel. Nearby extraterrestrials could quickly pop over for a neighborly visit, if only to level our cities.

Indeed, the concept of local company is so ingrained that "Martians" has become a common pseudonym for "aliens." But the brutal truth is that four decades of space exploration have failed to reveal any other occupants of the solar system, Martian or otherwise. If we ever do find living creatures on these worlds and moons, they're likely to be simpler than a salamander. Our quest is for something a little more interesting. Our hope is to uncover other sentient beings, other intelligence that has been cooked up in the 15 billion-year history of the universe. To do so, we have to broaden the search to the myriad pinpoints of light that beckon us from afar—from regions of space that are at least ten thousand times more distant than Pluto's dim haunts. We must turn our attention to the stars.

The many hundreds of stars visible to the naked eye on a clear evening only hint at the Galaxy's immense stellar kingdoms. A half-trillion stars make up the massive pinwheel that is the Milky Way. And the Milky

Way, our Galaxy, is only one of 50 billion galaxies within the range of our telescopes, each hosting its own complement of hundreds of billions of stars. Stars are commonplace, yes. But they are also remarkable. To make a sun requires the implosion of several billion billion billion tons of gas over tens of millions of years. The creation of a star is a feat beyond all possible human engineering. And yet it has happened often and everywhere.

As you shuffle along the ocean shore on a summer afternoon, reflect on this: there are more stars in the visible universe than all the grains of sand on all the beaches of Earth. As your eye catches the innumerable glints reflected from the tiny, faceted particles at your feet, and as you step over dunes and sand clumped against rocks and weed, imagine that each grain represents a part of the universe in which a star is breaking the deep, black cold of space. Every grain is proxy for a stellar realm where energy may be pouring onto orbiting planets, and where the subtle processes of life may have begun. Each doubles for a place where billions of thinking creatures may be organized into societies, playing out their existence unseen and unnoticed by us.

Most scientists believe that we share the universe with other beings. If you ask why, their answer usually boils down to the fact that they have walked in the sand. They are deeply aware of the staggering size of the cosmic arena.

The potential sites for other thinking beings are incredibly plentiful. Unfortunately, they are also extraordinarily remote. Unlike the grains of sand on a beach, the stars are separated by daunting voids. They are buffered by emptiness. Mankind has managed, with considerable effort, to send robotic probes throughout the solar system. Our rockets can manage a trip to Mars, even to Saturn or Pluto. But the stars are at distances that make solar system excursions seem trivial. Imagine our planetary system to be a village, with the Sun in the middle. In this model, Sol is a glowing sphere the size of a soccer ball. Mercury, Venus, Earth, and Mars, no bigger than small beads, are arrayed on the village's central square. Pluto orbits the town, and is hardly larger than a pinhead. But the nearest other star (which may or may not have a surrounding village) is a soccer ball four thousand miles away.

And that is only the nearest star. The array of soccer balls that comprise the Milky Way galaxy would, on this scale, stretch to the Sun. At 25 times this distance we would encounter the first of the nearer, large galaxies, with it own immense, flattened fields of soccer balls.

Travel to other stellar realms is quite beyond our capabilities, so we cannot search for intelligent aliens by simply going there. As a result, the hunt for cosmic company has shifted gears. We must now count on the possibility that evidence of the aliens' existence (not necessarily the aliens themselves) has, or will, come here. We must hope that distant societies are somehow making their presence known in ways we can recognize from afar. Perhaps the aliens are flashing lights or beaming radio waves in our direction, or engaging in unmistakable and easily visible feats of astro-engineering.

How should we approach the challenge of finding other beings at great distance? If the extraterrestrials are spread among the stars, is there some way to define the spread? Are some stellar habitats to be preferred, or is one star as good as another? Can we efficiently sift through the colossal abundance of cosmic sand to focus our attentions on the special grains that we seek?

SUITABLE STARS

When rockets revealed that thoughtful aliens are unlikely to be in the neighborhood, sci-fi writers responded by moving the homes of their extraterrestrial protagonists to the distant suburbs. In Steven Spielberg's blockbuster film, *E.T., The Extraterrestrial,* a pint-sized critter from the Andromeda Galaxy visits Earth to pick a few plants and, after being accidentally abandoned, to pass the time by playing with the kids. This premise clearly ignores the hard fact, just noted, that the distances between stars (let alone between galaxies) are stupefyingly great. But E.T.'s far-away habitat presents another puzzle. As we've mentioned, our Galaxy comprises a half-trillion stars. This is a quantity sufficiently impressive that it's a bit startling that E.T.'s place of residence is elsewhere, in another galaxy altogether. Why did E.T.'s search for interesting flora require such an immense voyage? Could it be that among the hordes of stellar habitats in an entire galaxy, there are none up to E.T.'s shrubbery standards?

That seems unlikely. The stars of the Milky Way come in a range of sizes, from dim and diminutive brown dwarfs to super-super giants hundreds of times the Sun's girth and tens of thousands of times its brightness. Of the stars charted by astronomers, those much larger than the Sun, while impressively hefty, make up only a small fraction, one percent of the total or so. But these biggest and brightest stand out. They comprise the bulk of the stars we can see with the naked eye, the familiar denizens of the night sky. Few are likely to host inhabited planets however, because large stars are either very young or very old.

Red giants (a familiar example is Betelgeuse, in the constellation Orion) are stars that have entered their golden years. After exhausting their natal supply of hydrogen fuel, these geriatric stars swell up to sing a final, impressive swan song. They huff and puff as a red giant for a few hundred million years before finally collapsing to a tiny stellar corpse. During this brief bulk-up, the dying stars swallow any planets that orbit them tightly, and bake those that orbit farther out. Incidentally, this is the inevitable fate of our own solar system, approximately five billion years hence. Our Sun will fatally inflate, consuming the Earth and crisping the outer planets. Our Sun and its attendant worlds are doomed. But in the meantime, as we scan the skies for aliens we can skip over the doddering, red giants. Any civilizations that may have existed around these bloated stars will have long ago been toasted or consumed.

Blue giants are stars that are much heavier than the Sun. They are stellar newborns, their interiors so tightly crushed by their exceptional mass that they burn with blue-white ferocity. These, too, are uninteresting hangouts for possible aliens, because stars that are born big, die young.

How young? Consider nearby Sirius, the brightest star in the night sky, easily seen southeast of Orion's familiar form. A super stellar luminary like Sirius will run through its supply of nuclear fuel in ten million years or so, and then ignominiously blow up in a massive explosion called a supernova. While the supernova will intrigue astronomers who have the opportunity to study it at great distances, for Sirian locals the star's dramatic end will be a catastrophe. The supernova will flash-sterilize any orbiting planets, destroying all life with a blast of deadly radiation.

In fact, this regrettable situation is of little concern, as there won't be any locals, at least none that are intelligent. Sirius rushes through its stellar life too fast. It lacks the patience required to foster the slow, biological processes that might ultimately produce thinking beings. After ten million years, a time too short for any attendant planets to incubate even simple life, this bright star commits suicide. Our own Sun, a star of medium size, has already lived hundreds of times longer. This is fortunate for the readers of this book, as it took more than four billion years of slow evolution before *Homo Sapiens* made his debut on Earth's stage. Stars that are born big and bright are like pop music stars: they burn out young. There is little incentive to look for alien societies orbiting these stellar behemoths.

At the other end of the stellar size scale are the dwarfs, stars that are half the Sun's bulk or less. Just as in nature, where small critters greatly outnumber large ones, small stars greatly outnumber large stars in the Galaxy (although, because they are so dim, relatively few can be seen without a telescope). For every Sun-sized star, there are ten dwarfs, more or less the ratio enjoyed by Snow White. But, while plentiful, these diminutive stellar orbs are less than attractive for hosting inhabited worlds.

Small stars have low energy output. Consequently, any encircling planets will receive little succor against the deadly cold of space. Most will orbit in a bleak and permanent state of killing frost. Only planets that tightly gird the star might be able to keep their oceans and topography above freezing. Unfortunately, the worlds that snuggle up to their bantam stellar hosts for a little warmth are susceptible to another lamentable affliction known as "tidal locking." When two astronomical bodies orbit one another at short range, gravity exerts a significantly stronger pull on their facing sides than on their far ones. The result is that the smaller of the two bodies soon becomes "tidally locked"—its rotation about its own axis is synchronous with its orbital rotation about its bigger celestial brother. In other words, it develops the condition in which one side always, or nearly always, faces its companion.

This is the situation for the moon, which orbits relatively near to Earth. A similar fate has also befallen Mercury, the closest planet to the Sun. Mercury spins exactly three times on its axis for every two rotations about the Sun. The consequence for this small planet is that one side is bathed in sunlight for 176 days, long enough for it to achieve the toasty temperature of 800 F. On the other side, temperatures are insufferably frigid. Tidal locking has ensured that Mercury is worthless for biology, although it might be a good place for a solar observatory or a tanning salon.

Imagine a planet orbiting close in to a dwarf star. Suppose that this world is big enough to have an atmosphere and water. Such a planet might at first appear to be an attractive place for E.T. But it won't be, because tidal locking will produce extremes of temperature provoking continuous, howling winds. In the movie *Alien,* the crew of the Nostromo is lured to a planet of incessant cloud and perpetual storm. This nightmare world is a good cipher for the inner worlds around small stars. While they constitute the overwhelming majority of the Galaxy's stellar population, dwarf stars are unlikely homes for worlds with sentient beings.

As far as E.T. is concerned, neither big stars nor dwarfs are appealing centerpieces for his native planet. Most stars get a thumbs down from the wrinkled extraterrestrial. But that still leaves some suitable candidates, roughly one star in twenty. Neither giant nor midget, this privileged minority consists of stars that are comparable in size and brightness to our Sun. They are the Sun's cousins. Since the total tally of stars in the Galaxy is approximately a half-trillion, the number of cousins is substantial: about 25 billion. The bottom line? It seems that only a small fraction of the stellar populace qualifies as a suitable sun for alien habitation. But there are so many stars to begin with, this qualifying minority is still massively plentiful. It seems that E.T. needn't have wandered so far afield. Our galaxy, and his, teems with stars that are suitable for hosting habitable worlds.

But do they? Are planets plentiful? Or are most stars surrounded only by sterile space?

WORTHWHILE WORLDS

Intelligent life, if it exists elsewhere, will have surely begun its existence on a planet capable of supporting life. The thin surface of a planet is where gaseous atmospheres and liquid water meet. These circumstances encourage fast chemistry, and life *is* chemistry.

How many of the billions of Sun-like stars in our Milky Way are accompanied by planets? Are such small, round worlds rare and precious, or a dime a dozen?

Until very recently, no telescope on Earth had a chance of actually *seeing* extra-solar planets. These far-flung worlds would be dim and distant. So astronomers resorted to attacking the question of their existence indirectly: by trying to understand how planets arise. If making planets is the result of circumstances common in the Galaxy, one can safely assume that a lot of planets whirl through its cavernous realms.

The formation of planets is an old problem in astronomy. Indeed, for most of the period since the Renaissance, the creation of the solar system (known as cosmogony) has been *the* problem for astronomy. In the 18th century, the German philosopher Immanuel Kant and the French astronomer-mathematician Pierre Simon de Laplace advanced the theory that the solar system came into being when a giant cloud of material collapsed into a flat disk. The Sun formed at the center, and planets condensed around it. Kant and Laplace found an encouraging, small-scale analogy for their idea in early telescopic views of Saturn's ring and moon system.

Their so-called "nebular hypothesis" fell into disrepute at the beginning of this century when astronomers, studying the spin and orbital motions of all bodies in the solar system, concluded that the Sun rotated too slowly to have been formed from the same stuff as its planetary entourage. (The Sun takes approximately 27 days to turn once). In the language of physics, most of the angular momentum of the solar system was carried by the planets. If Sun and planets had condensed out of a common birth cloud, the astronomers argued, the angular moment should have been more equally shared. So the scenario of collapsing clouds became *passé*.

A new and cataclysmic cosmogonic theory was concocted to replace it. As originally proposed by Chicago geologist T. C. Chamberlin, this theory held that sometime in the long-forgotten past, the Sun had a grazing collision with another star, a dramatic, slow-motion stellar fender-bender. While the two giant, glowing orbs approached and passed one another, gas was torn off their fiery surfaces. The interloper star then sailed slowly into the Galaxy's dark voids. But the ripped-out wreckage was left behind to cool and ultimately condense to form the planets.

This close-encounters-of-the-stellar-kind scheme was exciting and dramatic. It even made a gratifying appeal to biological genesis: two bodies come together, and the result is planetary progeny. Unfortunately, Chamberlin's spicy scenario violates the Principle of Mediocrity. It requires that we are the product of unusual circumstances. The near-collision of two stars is *extraordinarily* unlikely. Picture again our soccer ball Sun, in the village that is the solar system. It is moving in a random direction at about an inch or two an hour (and, incidentally, taking the whole village with it). What are the chances that it will ever brush up against another soccer ball, when the nearest of those is more than four thousand miles distant and is itself moving in a random direction at the same, slow speed? In the entire history of the Sun, such an encounter will have less than a one in a billion chance of happening. If this is truly the way that planets are formed, then we are extremely lucky to have one to occupy. Our situation is anything but mediocre, our solar system is wonderfully special, and we might very well be alone in the Galaxy.

Chamberlin's idea could have discouraged any search for cosmic company, except for the fact that planet formation by stellar sideswipe is no longer fashionable. Astronomers have found new explanations for the Sun's slow spin, and nebular theories, similar to what Laplace and Kant had proposed, are back in favor. The current formula for making planets goes something like this: begin with a giant molecular cloud, a massive clot of excruciatingly cold gas and dust. Such clouds pepper the dark reaches of our galaxy. Now arrange that, by accident, the cloud is disturbed, either by the nearby presence of another cloud (not an unlikely event) or by its passage across one of the Milky Way's spiral arms. The disturbance sends ripples through the cloud that trigger the collapse of small portions of its gas and dust interior. The central parts of these collapsing regions are soon under enormous pressure as gravity pulls atom atop atom. The gas heats up, and the internal temperature skyrockets to millions of degrees. Nuclear reactions begin, and a star is born.

Some of the collapsed material, which is invariably slightly rotating, gathers into a disk, encircling the newborn star like a giant hat brim. The molecules of the disk occasionally stick to one another, and form small grains. These stick to other grains, and grow, much like rolling snowballs. Soon, an assortment of clotted gravel and chunks of ice float about the young star. Some of these nascent planetesimals eventually reach the size of a small mountain. This is a critical juncture, for once a mile or so across, their gravitational attraction begins to actively solicit new material. The bigger planetesimals swell quickly as their increasing size causes them to engorge their smaller brethren. As material continues to slam into these would-be worlds, it heats them to a sticky, molten state. For the largest objects, self-gravity will be sufficient to pull the hot material into the most compact arrangement possible, a sphere. Smaller bodies will end up with somewhat irregular shapes, like potatoes. In the end, the condensed remnants from the birth of a star have become planets, moons, asteroids and comets. They are distinct bodies, embarking on a perpetual lifetime of servitude orbiting their newly-formed stellar master. Another solar system is added to the Galaxy's tally.

All of this sounds easy, and maybe it is. But there's a time limit for it to happen. While grains are graduating to mountains, the stellar newborn at the center of the action is graduating from dim fetus to shining star. Once the star gets its nuclear furnaces sufficiently stoked and burning, it will rapidly blow away any encircling material that hasn't agglomerated into large chunks. Only if the encircling material manages to form heavy-duty planetesimals before this housecleaning takes place, will the young star be left with a retinue of worlds.

This modern theory of solar system formation suggests that essentially all stars get a chance to make planets, although not all may succeed. But theories clearly come and go. The ultimate arbiter of all astronomical

A model simulating the process of planet formation.

theory is always the telescope. What do we see when we probe nearby stars? Do they indeed sport planets?

Until rather recently, no one knew the answer to this question. It is only since 1993 that astronomers have found compelling evidence for *any* planets other than the familiar nine of our own solar system. To discover the evidence isn't easy. Planets shine by reflected light, and are typically at least a billion times fainter than their host stars. Imagine a Hollywood searchlight being buzzed by a moth. Observing the flitting moth from ten thousand miles away is analogous to the astronomer's problem of seeing a planet orbiting another star. A planet's dim glow, confounded by stellar glare, is so difficult to detect that even the refurbished Hubble Telescope cannot do it.

And yet, planets *have* been found, their existence revealed by indirect methods. Instead of looking for the moth, planet hunters search for its effect on the searchlight. The presence of unseen worlds is inferred from the slight gravitational tug they exert on their stellar masters.

This works as follows. Consider a star hosting a single, massive planet.

The common view of solar system kinematics, and one that a visit to the local planetarium will mistakenly reinforce, is that planets parade in near-circular (actually, elliptical) orbits about a fixed sun. The truth is slightly different. Both the star and its planets orbit their common center of mass, a shifting mathematical point that lies between them. Imagine that the star and planet are sitting in your spacious laboratory, and have been connected by a long stick. Further imagine that you succeed in balancing that stick on a nail. The point of balance is the center of mass, and both star and planet will rotate about this point. Of course, since the star is much heavier than the planet, the center of mass will be much closer to the star, typically only a half-million miles or so from its fiery center. So while the planet travels in a big orbit, the star travels far more sedately, in a small, tight orbit. Nonetheless, that modest stellar wobble can be detected with a telescope at great distance because the star is putting out so much light.

Stellar wobble is a tip-off to planetary companions. If stars are seen to oscillate back and forth on the sky or ping-pong toward and away from us, we can infer that they are ringed by orbiting bodies. The size of the unseen worlds can be determined from the speed and vigor of the dance. The gratifying and good thing about this approach is that, despite being indirect, it is the only one to have succeeded so far in finding planets. Indeed, it succeeded even *before* it succeeded.

Ever since the late 1930s, astronomers have used the 24-inch telescope of the Sproul Observatory, near Philadelphia, to monitor Barnard's Star, a dim bulb a thousand times fainter than the Sun. Barnard's Star is a mere 5.9 light-years distant, making it our next-to-nearest stellar neighbor (the Alpha Centauri system is somewhat closer at 4.5 light-years). Photographs taken during the course of many years seemed to show a side-to-side wobble for this close-by dwarf. The minuscule wobbles revealed themselves when astronomers used a microscope-like device to compare the position of Barnard's Star in photos taken days and weeks apart to that of other, more distant stars. The device essentially allows the researchers to stack two photos so that all the distant background stars line up. They then carefully examine whether the star being scrutinized (Barnard's Star, in this case) also lines up. It didn't.

In the 1960s, after two dozen years of observation, astronomer Peter van de Kamp announced that two Jupiter-sized companions were responsible for Barnard's Star's apparent shaking. His remarkable claim was that this nearby star—indeed, a *very* nearby star—had planets. The result augured well for a Galaxy filled with small, potentially habitable worlds.

Unfortunately, van de Kamp's result was bogus. In the early 1970s, astronomers using other telescopes, the 20-inch refractor at the Van Vleck Observatory and the 30-inch instrument at Allegheny Observatory, pointed out that they couldn't find Barnard's Star's intriguing oscillations. Since then, researchers have concluded that the wobbles reported by Van de Kamp were likely due to lens adjustments occasionally made at the Sproul Observatory. The shaking was taking place in Pennsylvania, not 5.9 light-years away.

Despite the false alarm, and to some extent mistakenly encouraged by it, astronomers persisted. They pressed the search for visible wobbling of nearby stars. At the same time, other research teams refined techniques for sensing to-and-fro motions.

Model of a dead planet orbiting a pulsar.

In 1992, radio astronomers Alex Wolszczan and Dale Frail were busy studying pulsars using the giant Arecibo Radio Telescope in Puerto Rico. Pulsars are the spinning, crushed corpses of defunct, giant stars. They beam light and radio waves into space with exacting regularity. Wolszczan and Frail were listening to the rhythmic radio beats of the pulsar PSR 1257+12[2], beats that occur 162 times a second. But the two astronomers found that the radio flashes sometimes arrived a thousandth of a second earlier or later than expected. Over the course of months, PSR 1257+12 apparently slows down or speeds up. The ready explanation for this slight arrhythmia is that the pulsar is moving to and fro—wobbling ever so gently.

Wolszczan's analysis of the wobbles implies that at least three planets journey around the pulsar at distances that are similar to the orbits of Mercury and Venus. These unseen bodies are lightweights, Earth-sized and smaller. Because of their ability to clock the pulsars with stupendous precision, Wolszczan and co-workers could measure wobbles as subtle as two feet per second, the speed of a crawling baby. The pulsar crowd could, in principle, detect a planet no more massive than the moon[3].

However, despite their attractive size and orbital distance, planets around pulsars are non-starters in the race to produce intelligent life. Pulsars are the imploded remnants of a large star, one that has committed youthful hara-kiri as a supernova. The orbiting worlds inferred to exist by the radio astronomers are captives of a dangerous corpse, strobed by high-energy beams rather than a benevolent, warming glow. PSR 1257+12 is a pathological system, and an unlikely home for life, intelligent or otherwise.

---

2. Note that pulsar names are simply the coordinates of their positions on the sky. This is a handy, if unromantic nomenclature, and similar to the frequent European custom of adopting surnames that are the name of the town in which you live.

3. A recent result by astronomers Klaus Scherer, Horst Fichtner, John D. Anderson and Eunice L. Lau suggests that the innermost and least massive planet claimed by Wolszczan may be spurious. The new result attributes some of the smaller wobbles observed in 1257+12 to a very local effect, irregularities in the stream of charged particles boiling off the Sun known as the solar wind. The solar wind wafts through interplanetary space, and slightly affects the arrival time of pulses from 1257+12.

The discovery of planets in such oddball circumstances further encouraged those who were hoping to find new worlds in more conventional haunts. Astronomers using optical telescopes accelerated their search for any to-and-fro shaking by ordinary stars. The tools of this trade are spectrographs, not photographs. A spectrograph, which in its most elementary incarnation is nothing more than a prism, spreads the light from a star into its component colors, or wavelengths. Fortunately for astronomers, the constituent atoms of a star's outer layers modify the light pouring from the inferno below by absorbing certain wavelengths. These dark features in the star's spectrum are easily seen, and have become well-known markers. By measuring the precise wavelengths at which the atomic imprints occur, astronomers can deduce the speed at which the star is moving either toward or away from us. If, like the sound of an approaching train's whistle, the wavelengths are shortened (higher pitch), then the star is moving our way. A shift to longer wavelengths (lower pitch) implies a receding star.

The contemporary ability to measure very small spectral shifts means that researchers can now sense the to-and-fro dance of a distant star even if it moves only a few tens of miles per hour. This is the amount of motion that Jupiter provokes from our own Sun.

In 1995, two Swiss astronomers, Michael Meyor and Didier Queloz, made a discovery that will still be set down in astronomy textbooks a thousand years from now. Meyor and Queloz were studying local stars in the hope of finding wobbles. But their prey was not planets. Rather, they wanted to learn if unseen and diminutive stellar companions—so-called "brown dwarfs"—were prancing about their target stars. Intermediate in size between planets and small stars, brown dwarfs can't muster enough pressure to coax nuclear burning from their gaseous innards. They glow only dimly, from the leftover heat of their birth, and consequently are extremely hard to see directly.

The two Swiss astronomers were delighted to find that the Sun-like star 51 Pegasi showed unmistakable to-and-fro motions of roughly 110 miles per hour. But when they worked out the mass and distance of 51 Pegasi's invisible dancing partner, they were astounded to learn that it weighed in at no more than two Jupiters. The star's unseen companion

Model of the Jupiter-sized planet recently detected in orbit around 51 Pegasi.

was obviously too small to be a brown dwarf. It was planet-sized
—the first planet to be detected around an ordinary star.

While 51 Pegasi is a run-of-the-mill sun, the new world discovered by
Meyor and Queloz is not. It circumnavigates 51 Pegasi in a mere four
days (against 88 days for Mercury's trip around the Sun and 365 days
for Earth's). This blistering speed is a consequence of the planet's tight
orbit. Meyor and Queloz's world is so close to its host star that temperatures are reckoned to be 2,500 F, or the melting point of cast iron.
Hot and heavy, 51 Pegasi's superheated planet might be nice to study,
but neither you nor E.T. would want to live there.

The discovery of a world orbiting another sun was, like Columbus'
discovery of America, unexpected. It was, as are so many astronomical
discoveries, serendipitous. Other research groups had spent years vainly
scrutinizing stars for the delicate motions that would betray planets.
In some cases, the astronomers involved in these efforts were simply
unlucky; their observing lists were too short to have a good chance of
including a star with a massive companion. Other astronomers hadn't
anticipated that any large planet would embrace its host star so tightly,

and were looking for wobbles that spanned years, not days. Until Meyor and Queloz, all attempts to find the elusive dance of the stars had failed.

It is also ironic that the 51 Pegasi planet was found by workers in a country not especially noted for astronomical discovery. The world's stellar research heavyweights—the United States, Great Britain, France, Holland, Italy—are generally countries with long naval histories. Astronomy, after all, once served very practical ends: navigation at sea. But the Swiss are not a maritime power. Their country has no coast-lines, and while Swiss Army knives abound, there is no equivalent cutlery from the Swiss Navy. When asked what tradition had fostered astronomical study in his native, land-locked country, Michael Meyor opined that it was the watch industry: "The Swiss watchmakers need-ed good astronomical time keeping to check their product. So a few observatories were built. But of course, making measurements to check the time is boring, and only takes a small effort each day. The astronomers at these observatories soon turned their attentions to research."

Even though the pulsar planets were intriguing, they didn't electrify both astronomers and the lay public the way Meyor and Queloz's dis-covery of a world around a *normal* star did. The latter was, after all, the type of system people wanted to find, and the discovery was soon followed by a torrent of announcements that other stars also sported planets. With impressive regularity, these announcements came from a talented pair of planet hunters wielding the Lick Observatory's 120-inch telescope near San Jose, California. Astronomers Geoff Marcy and Paul Butler had been stunned by the Swiss result. The California duo had long been searching for planet-induced wobbles around nearby stars, and the Swiss announcement caught them unawares. So when news of 51 Pegasi hit the streets, Marcy and Butler felt scooped. And they were, of course. Unless Mayor and Queloz were wrong.

Marcy and Butler checked it out. They swung the Lick telescope in 51 Pegasi's direction. Within days, they knew the truth: the wobbles were there and the Swiss were right. The Californians could no longer be the discoverers of the first planet around a Sun-like star. But with their own research now kicked into high gear, they soon began churning out planet discoveries at the rate of one a month. The worlds they discovered were, like 51 Pegasi, hefty. They were frequently Jupiter-sized or larger, and mostly in close orbits (although none so close as 51 Pegasi). The bias towards massive planets was the inevitable consequence of their observing scheme: Marcy and Butler were looking for to-and-fro wobbles. And close-in, heavy planets make the largest such wobbles.

A year after Meyor and Queloz changed the astronomical landscape with evidence for worlds around nearby, normal stars, there were more planets known outside the solar system than in it. Geoff Marcy had predicted that continued observations would guarantee a "rain of new planets." By 1996, the storm was well underway.

Jupiter as photographed by the Hubble telescope in October 1991. [NASA]

GOOD PLANETS OR BAD?

The initial hunt for planets has scored success in a few percent of the stars that were searched. But the score has been held artificially low by instruments that can only detect wobbles for stars harried by large, close-in planets. The actual fraction of stars ringed by orbiting worlds is surely higher than the astronomers' two or three percent hit rate. It could be 50% or more. Even if we accept a far more modest figure, say 5%, the number of solar systems in our galaxy will tally in the billions.

That's the good news. Those seeking intelligent life elsewhere have spent decades wondering whether planets were precious and rare, or a dime a dozen. Now we know the approximate price, and it's cheap. The bad news is that we don't know if *habitable* planets are common. Jupiter-like worlds, such as those discovered by the wobble watchers, are thought to be born in the outer, icy regions of a young star's nebular birthplace. They will be swathed in choking clouds of industrial-strength ammonia and methane. While dramatically impressive, they are not appealing habitats. However, it's no great conceptual leap to presume that where large worlds exist, small ones do as well. And it is on small, rocky planets that the salubrious combination of atmosphere and liquid water can produce the conditions we believe will incubate life.

Establishing that small, Earth-sized planets really do pepper the cosmos has become a research holy grail. By pushing their technique using the much larger Keck Telescope in Hawaii, Marcy and Butler will be able to track down planets smaller than the massive companion to 51 Pegasi. They expect to discover worlds the size of Uranus and Neptune, a small step in the right direction. Meanwhile, NASA has announced a new research initiative, the so-called "Origins" program, to hunt down Earthly abodes around other stars. Most of the proposed schemes for finding these distant, terrestrial globes involve putting telescopes in space, and it will be a decade or more before such plans are realized. Until then, we can only speculate that the small planets are out there.

Size is important, but as Dr. Ruth would caution, it's not everything. Location plays a big role. A planet with pretensions to habitability must be located—by definition—in its host star's "habitable zone." This is most conventionally defined as that range of orbital distances within which an approximately earth-sized world can sport liquid water. The inner edge of the habitable zone is where a planet's oceans would inevitably boil away, and the outer boundary is marked by permanent freezing of all water.

Consider the habitable zone of our own solar system. Obviously, Earth is situated within it. But how far does the zone extend inward and outward? Venus, which receives nearly twice as much sunlight as the Earth, is a world where the greenhouse effect has run rampant, leaving the planet without water, and most assuredly without life. It is clearly beyond the tolerable pale. A conservative view would hold that the inner edge of the habitable zone is not much closer to the Sun than is the Earth. The outer boundary probably extends slightly beyond Mars, whose frigid physiognomy is due more to its puny size than to its distance from the Sun. Consequently, we can say that two of the nine planets in our own solar system are in the habitable zone (and we haven't even considered in this discussion such far-out locales as the moons Europa and Titan). If this rather generous fraction is typical, then most *other* solar systems will have at least one world that biology could bless.

All of this is to suggest that earth-like worlds could be as common as flies. This upbeat conclusion is based on both fact and speculation. The *fact* is that recent discoveries have shown that planets are plentiful. The *speculation* is that many of these are earth-like and in solar systems similar to ours. But some researchers are uncomfortable with this easy optimism. Speculation that generalizes from the situation of our own solar system (as we have repeatedly done) is, in their view, obviously suspect. We impose the conditions of our home planetary system on putative other systems, calling such conditions "normal," because we are here to do so. In a solar system that winds up sterile or devoid of intelligence, there are no pundits making speculations. For the man born in a castle, castles are the standard habitation. Our view could be heavily biased.

This is an obvious challenge to the sacrosanct Principle of Mediocrity.

We like to claim that we're not special; our solar system is garden variety and generic. But could it be that we've protested our modesty a little too much? For example, in his book *What if the Moon Didn't Exist?* Neil Comins notes that Earth is unlike its fellow rocky worlds (Mercury, Venus, and Mars) in that we have an improbably large moon. Mercury and Venus have no moons, and the two moons of Mars are small enough that they could be parked within the confines of downtown Los Angeles. Earth is blessed with a relatively hefty satellite. Without this moon, according to Comins, the resulting "lower tides, higher winds, and shorter days [would make our planet] a much less hospitable place in which to live." A more extreme position is that, since the moon has stabilized the Earth's axial tilt, preventing large, rapid changes—changes that would induce massive and sudden alterations of climate—our very presence on Earth is contingent upon this massive, natural satellite. Without the moon, we wouldn't be here. But the moon's existence could be just a rare bit of luck. Our satellite was probably created by a cataclysmic collision of Earth with a Mars-sized asteroid nearly four billion years ago. The fact that we have a lunar-sized companion is apparently due to a fortunate accident. This suggests that Earth's situation could be rare in the Galaxy.

Not so, responds Luke Dones, an astronomer at NASA's Ames Research Center, in Mountain View, California. A new-born solar system will be like a shooting gallery, as nascent planets careen hither and thither. In such a maelstrom, the formation of occasional, massive moons by collision might be a common event, and not the least bit rare. Additionally, even a moonless world whose rotation axis was frequently awry might still be habitable if it had large oceans to moderate the consequent temperature imbalances.

Other examples of how Earth might be special have been offered. We are fortunate, it has been said, that our planet spins fairly rapidly, probably the result of those early collisions with asteroids. It was once thought that any world taking longer than a hundred hours to turn on its axis would be a fourth-rate planet for life. The days would be too hot, and the nights too cold. But once again, Dones argues that such pessimism might be unfounded. Big oceans or thick atmospheres could ameliorate conditions on such lazy planets.

Of course, while a collision now and then between asteroid and planet early in the young world's life might have the beneficial effects of giving it a healthy spin and supplying it with a large moon, too much of this good thing would be a *bad* thing. The dinosaurs and 90% of all Earth's species were wiped out by the large rock that slammed into the Yucatan 65 million years ago. If hordes of asteroids cruised the solar system now, such catastrophes would happen so often that intelligent life might never gain a foothold. Fortunately, Jupiter has long played the role of peacekeeper, deflecting most of these dangerous leftovers from the solar system's creation into its far reaches, where they present little danger. Without Jupiter to clean out bullies from the inner solar system, the reader of this book might not exist. But are we special in having a massive planet to perform this critical chore? Once again, the answer is "probably not." Modern cosmogonic theory envisions that large planets will be a natural and common consequence of star formation, and the discoveries of the planet hunters show that massive planets exist in abundance. Nearly everyone will have a Jupiter.

But this last point could be a double-edged sword. Big planets dominate the list of newly-discovered worlds, behemoths that orbit near to their suns. This circumstance might conceivably be a point of argument for the skeptics, for those who feel that Earth really does enjoy special astronomical amenities. Our current ideas about planetary formation hold that small, rocky planets will form close to newborn stars, and large, gassy Jupiters will develop on the outskirts. This is obviously the situation in our own solar system, where Mercury, Venus, Earth and Mars populate the inner realms, and Jupiter, Saturn, Uranus and Neptune reign in the suburbs. It came about because radiation from the nascent Sun drove light, gaseous elements out of the central part of the collapsed disk of material forming the planets. The result of this cleansing process was that the icy compounds that make up Jupiter were largely relegated to the outer reaches of our solar system. But most of the extra-solar planets discovered since 1995 are like 51 Pegasi, large and in small, tight orbits. It seems that for these solar systems, a sticky fog of left-over disk material acted to drag all the new-born worlds inward. Any small, rocky and habitable planets were presumably swallowed, slowly spiraling into the maws of their stellar masters. What they left behind were the unappealing Jupiters.

If such is the usual fate of planetary systems, we will be forced to conclude that Earth is a galactic specialty item, rare and possibly alone. The Principle of Mediocrity will be stopped at the door of our solar system. However, there is simply no warrant to jump to this sobering conclusion. Remember that the spectrographs used to find stellar wobbles are currently incapable of uncovering anything other than close-in, massive planets. Look at it this way: we have scanned Africa from the air, and have seen only elephants. But a good pair of binoculars would reveal the myriad small animals that are beyond the capabilities of a first reconnaissance. Similarly, we have only begun to scout for other planets. During the next decades astronomers will set into play instruments capable of finding small, Earth-like planets. Few anticipate failure.

A HOME FOR E.T.

When wrinkled little E.T. finally had enough of playing with the kids, he pointed a spindly finger to the sky and declared his intention to "phone home." The visiting alien didn't specify exactly where "home" was (other than to note that it's in the Andromeda galaxy), but we all assume E.T.'s big-eyed brethren live on a planet. A planet not very dissimilar to Earth.

Planets are everyone's favorite residence, at least among those who speculate on where the extraterrestrials might hang out. Of all the real estate in the universe, it is only on planets that liquid water, a gaseous atmosphere, and the crusty surfaces of continent and sea bottom can set in motion the still-mysterious chemical reactions that lead to self-replicating, growing systems. Yes, it's possible that alien intelligence has spread beyond the limited confines of planets, and we will return to this idea later. But no matter what turf the extraterrestrials now claim, it seems strongly probable that they got their start on the cool, solid worlds where chemistry most readily occurs.

The bottom line is simple: if there are no decent planets in the teeming star fields of space, then there aren't likely to be any extraterrestrials either. E.T's evolutionary line will have required a planetary home for at least some period in its history. For thousands of years, the existence of planets around other stars was the province of speculation and theory. But the recently measured, minuscule shakings of some nearby stars have brought home the hard evidence that our solar system is merely one among many, one among billions. Many of these newly-discovered worlds orbit sun-like stars, the very stellar environments that we believe are the most beneficent for incubating advanced life. Earth-like worlds have yet to be found, but that is most plausibly understood as a consequence of our still-limited ability to find *any* planets. The existence of small planets is not seriously in doubt, and it seems more than likely that the daunting spaces of the Galaxy are filled with countless planetary homes.

If so, what sort of inhabitants might move across their distant, unseen landscapes?

How They're Built

In 1977, researchers aboard the three-man submersible *Alvin* were plumbing the Galapagos Rift, a cleft in the sea bottom off Ecuador's coast. More than a mile below the surface, conditions in the Rift are unspeakably harsh. No light penetrates its inky depths, and the pressures are enormous. The fissured ocean floor is punctuated by natural chimneys called "smokers," undersea vents that spew forth sulfurous, scalding waters. It is a liquid version of hell. Peering through Alvin's small windows, the researchers were surprised to find that even this dark and dismal environment was inhabited. Long, whitish tube worms, capped by red, velvety heads, swarmed around the churning smokers. The tube worms are bizarre animals, with neither mouth nor gut, living off the chemical processes of the bacteria that fill their innards. No one could have predicted the appearance of these exotic creatures.

What about the inhabitants of other planets? Can we say anything about E.T.'s construction? Will he be soft and squishy, hard and muscular, or clad in an insect-like carapace? Will he be the size of an ant or an aircraft carrier? Will he have two eyes, ten eyes, or simply a head swathed in waving, velvety antennae?

As we've seen, there is a growing line of suggestive evidence that our galaxy contains billions of Earth-like worlds. We dare to hope that life will have taken hold on many, and this expectation has been bolstered by the recent analyses of a Martian meteorite. So in considering what E.T. might be like, the first matter of inquiry is to ask whether biology is, indeed, rampant. Does the Galaxy teem with life? If so, would any of it be life-as-we-know-it, biochemically similar to the metabolic carpet that bedecks the Earth?

And what if it turns out that it is? If alien biology is both common and comparable to our own, then we might be tempted to assume that, given enough time, it would also produce creatures comparable to us, capable of both understanding life's process and seeking it elsewhere. Cleverness may inevitably appear on the scene. If this occurs on even a small fraction of the planets where biology has sprung up, we will have plenty of interesting cosmic company with which to share the universe. But, as we'll see, while most scientists suspect that life is widespread, there are some who feel that *intelligent* life might not be.

LIFE: MIRACULOUS OR NOTHING SPECIAL?

Where does life come from, anyway? The question is hardly new. One intriguing possibility is that it materializes out of nowhere. The idea of spontaneous generation, that life can quickly and routinely appear from non-living matter, was proposed by Aristotle more than two millennia ago. Like many faulty Greek scientific theories, it had a long run. Casual observation seemed to support it, after all, as anyone who has noticed the distasteful appearance of maggots in dead animals can attest. Louis Pasteur finally put the idea of instant life to rest in the last century when he noted that sterilized solutions would not spawn bacteria unless they were exposed to the air. But until then most researchers believed that even complex organisms could arise either from the usual dirty, nasty business of mating, or by springing forth without provocation from bits of matter, usually dead matter.

For those searching for extraterrestrials, the production of creatures via spontaneous generation would, quite literally, make life easier. Sure, it has disquieting side effects: new pets or house guests that unexpectedly appear in the basement like fire breaking out amid a pile of oily rags. But spontaneous generation would encourage us to expect life on any planet with a suitable pile of detritus.

Of course, life from junk is bunk. Biologists have replaced the theory of spontaneous generation with biogenesis. Life is produced by life. Maggots spring from tiny eggs that have been laid by living insects. But saying that biology produces more biology merely pushes our problem—the question of life's origins—back in time. Once life gets underway, it generates more. But how easy is it to get underway?

In 1936, Russian biologist Alexander Oparin published a landmark book, *Origin of Life,* that for the first time seriously postulated that life might have arisen on Earth, not by divine intervention, but by simple chemistry. This idea was later elaborated by his British colleague J. B. S. Haldane, who described the watery incubator of this first, primitive bit of biology as a hot, dilute soup. Oparin and Haldane's ideas became popularly known as the primordial soup theory.

How might broth beget life? If we can find an answer, it could have universal application. After all, if the path from porridge to protoplasm is not enormously difficult, then we might have a crucial insight into the genesis of E.T.'s early ancestors. We might be able to generalize from our situation on Earth to the myriad worlds that we expect are like Earth. If life is easy to get started here, it has surely sprung up elsewhere.

In pondering this idea, we should first have some inkling of what constitutes life. From a structural point of view, even the simplest, single-celled life is a rather sophisticated construction. It begins with a few elemental building blocks, the carbon, hydrogen, nitrogen, oxygen, phosphorous, and sulfur that make up 96% of the material of a living cell. This material is not rare. We know from the spectral analysis of starlight that these elements are abundant not only in the local neighborhood, but in the most distant corners of the Galaxy.

At the next level of complexity are the amino acids, simple organic molecules of which twenty are found in earthly life. (Chemistry allows thousands of amino acids, but nature, like your local fast-food joint, is parsimonious. It produces a wide variety of products from a small, select group of cheap ingredients.)

In a cell, amino acids are hooked together in long polymer chains to make the most common organic compounds of life: proteins. There are typically hundreds to thousands of amino acids lined up, conga style, in each protein, and several thousand different kinds of proteins in a single cell. Proteins are catalysts for the chemical reactions that are the basic business of life; they are responsible for metabolism.

The final, critical ingredients of life are the nucleic acids, DNA and RNA, also put together from some simple organic bases known, not surprisingly, as nucleotides. Familiar, helix-wound DNA and RNA are the blueprints for life, the mechanism for biological inheritance. Inheritance is absolutely critical since intelligent critters are perforce complex critters. Any beings that enjoy a more interesting lifestyle than just floating around in the muck must be the result of a long period of Darwinian evolution. They must have a device to ensure inheritance, a building plan that can be accurately passed on to new generations. Otherwise, progeny would be randomly different from the parent, and any improvements will be immediately lost.

In 1953, Stanley Miller, a student of University of Chicago chemist Harold Urey, conducted a simple experiment to investigate how much of the complicated edifice of life could be constructed by putting the elemental building blocks together and giving them a good "shake." Miller filled a flask with methane, hydrogen, ammonia and water vapor—all believed to be important ingredients of Earth's atmosphere four billion years ago. He then replicated the conditions of our primeval planet by subjecting the model atmosphere to heat and electric sparks (an analog for lighting bolts). After a week of cooking, Miller and Urey found that their apparatus contained two amino acids, glycine and alanine. It was like putting sand, concrete, fly ash, and water into a mixer for a week, and seeing cinder blocks drop out. This astounding experiment has been repeated several times, often using somewhat different energy sources and basic ingredients. The follow-up experiments also produced the compounds that underpin life.

The next step, turning amino acids into proteins, is not especially difficult. These important molecules of life were probably produced by naturally occurring reactions in the Earth's early atmosphere. Rain would drive the proteins from this airborne chemical factory down to the surface, where they could accumulate in tidal pools, mud flats, and deep in the ocean. After millions of years, they would constitute a vast food supply, awaiting a microbial dinner guest.

Exactly how that first biological diner was created is the mystery that has puzzled biochemists for decades. Obviously, making the amino acids is not the problem, as the Miller-Urey experiment showed. Connecting them into long chains as proteins is also easy enough. But at some point, the proteins need to be organized to form a cell, separating themselves from the chaotic outside world with a wall or membrane. They must be capable of both metabolism and reproduction. So what caused all these raw materials to self-assemble, to manage the step from proteins to protists, the simplest living cells?

We still don't know. The production of the nucleic acids, DNA and RNA, is a particularly difficult part of the problem. Astronomer Michael Hart once estimated that the typical wait before a molecule of DNA will randomly form in primordial soup is 100 thousand billion billion billion billion years, a length of time that dwarfs the age of the universe. Clearly, some set of circumstances must have existed on Earth to encourage, and considerably speed, the development of the Adam cell. We require some mechanism that will do (only) once what the now-discredited process of spontaneous generation was thought to accomplish every day of the week: the production of living matter from non-living molecules. If no such process exists, and Michael Hart is right, then life on this planet is an incredibly rare accident, comparable to winning the lottery jackpot on the first try. In that case, Earth could be the only life-bearing planet in an enormous, sterile universe.

Few people are keen to embrace this pessimistic view, and various theories pertaining to the origin of life, as this research topic is modestly called, are under consideration. Some biologists still envision the first microbes getting their act, and their morphology, together in the tidal pools so favored by the original soup enthusiasts. Somehow the proteins managed to organize themselves into a structure that could be reproduced almost without error. These replicating molecular assemblies ultimately evolved into the first cells.

Another point of view is that the first cells were born deep in the sea, in the hidden trenches where tube worms now wave their velvety heads. The sulfurous, boiling water spewing from vents in the trench provides an energy source where the Sun never shines. Remarkably, the present-day bacteria that share these dusky, hot seas with the worms are the most archaic life forms on Earth. The DNA in these single-celled denizens of the deep is grandparent to all other life forms. So if deep sea vents are where the most primitive life is found, an obvious possibility is that this is where life began.

It's a neat argument, but it's not conclusive. In fact, it's possible that these deep-dwelling bacteria are the oldest on the planet only because of an ancient catastrophe. Four billion years ago, when the Earth was still being pummeled by asteroids, a particularly large impact could have boiled away most of the oceans, killing everything except the simple life living in these watery trenches. We know such impacts occurred: the moon is the consequence of a particularly dramatic collision. If such a smashup occurred once life was underway, the bacteria of the trenches would have lived through it, protected in their deep-sea bomb shelters. We, and every other creature, are their descendants. That part is for sure. But if the impact scenario is valid, then these microscopic critters were not the progenitors of life on our planet, not the first living things, but merely the only survivors of a dramatic cataclysm. We can trace the family tree of life no farther back than these trench microbes, but possibly there are even earlier ancestors forever beyond discovery.

There are other possibilities, too. Perhaps the first cells began in clays, or as a consequence of the natural ordering of molecules in crystals. But while all such schemes are plausible, none has yet been convincingly demonstrated.

Part of the reason that the problem of life's origin is so difficult may be because we're looking at the latest models, the critters of today. Even the archaic bacteria of the deep sea vents might be substantially changed from their ancient ancestors. Examining a Stealth fighter will not give you tremendous insight into how the Wright Brothers designed their first flying machine. We may be having the same trouble divining the structure of the first cell. The technology of life has been refined, and in ways that are difficult to know. The first living things may not have had even short strands of that oh-so-complicated DNA. They may have passed on inheritance via some other mechanism that, during the course of eons, was supplanted and disappeared. The soft substance of cells leaves few fossils for our edification.

There's another, more radical suggestion for how biology got started on Earth: namely that it didn't. At the turn of the century, the Swedish chemist Svante Arrhenius proposed that spores from a distant planet could have managed to make an epic voyage across space to seed our planet. Life came from the cosmos. Arrhenius' hypothesis is dubbed "panspermia." (The terminology may sound sexist, but in Nature sperms usually travel more widely than eggs.) Of course, the panspermia hypothesis only obviates the necessity for cooking up life on Earth by shifting the burden of genesis elsewhere. Note that this idea isn't the same as the suggestion that Martian microbes might have infected Earth (or vice versa). A rock hurled from Mars could reach our planet in only a few years. But spores from a distant solar system would take *hundreds of thousands* of years to make the journey to Earth. They might not survive the trip. Interstellar space is brutally cold, and laced with deadly cosmic rays. Life, even as comatose spores, might have a hard time remaining lively in a space journey that extends over many millennia.

Francis Crick, famous as one of the two researchers who first unraveled the secrets of DNA, has suggested that this weak point in the fertile-rain-from-space hypothesis could be remedied by giving the spores protected passage on one of E.T.'s interstellar spaceships. Directed panspermia, as this Johnny Appleseed view of genesis is called, would mean that it was E.T. who deliberately brought life to Earth. Presumably, he's done other planets a similar favor. Not only do we share the universe, but everyone in it is a distant relative.

Panspermia, while appealing to some researchers, is rejected by most. After all, it doesn't really solve the problem of getting life started, but merely suggests that it didn't have to happen too often. In addition, you might ask where the sperms from space are today. Why don't they infect open Petri dishes? It's also interesting to note that if panspermia is right, if the Galaxy is inhabited by multitudes of distant relatives, then we might be vulnerable to universal diseases. Whole sections of the Milky Way could be knocked out by viruses spawned on a distant planet long ago.

At this point the reader might wonder whether to be optimistic or otherwise that alien worlds could be crawling with life. Life is a complex edifice, and while we're quite certain that nature will quickly produce abundant bricks, some assembly is still required. That assembly process has yet to be understood. But there's a very simple argument regarding life's origins that short-circuits virtually all of our ignorance and much of the preceding discussion. The argument derives from a solid observational fact: the oldest fossils of microbial life on Earth date from 3.6 billion years ago or more. Fossil evidence reported in 1996 suggests that life had a toehold on Earth 3.9 billion years ago. In other words, the oldest fossils are about the same age as the oldest rocks! (It's quite hard to find remains of life much older than these examples, even if they exist, because virtually all of the ancient rocks on Earth have been ground up, melted, or otherwise abused by tectonic activity.)

The age of the oldest microfossils is highly significant. Though no media folk were around to report it, 3.9 billion years ago our planet underwent an important change: it became a kinder, gentler place. The storm of asteroids and comets that had bombarded the planet abated. Earth's surface cooled, and rain could finally fall and collect as oceans. The fossil record indicates that, within a few hundred million years (at most) after the Earth became habitable, it was inhabited. A hundred million years is, in geological terms, only a few ticks of the cosmic clock. We may not understand how the first cell came into being, but Nature was clearly better at doing biochemistry than the biochemists. Life formed as soon as it could, and this remarkable fact is a compelling reason to believe that it will spring up on any half-decent planet. The froth of life will carpet a billion worlds or more in our Galaxy. This point of view isn't changed if, for reasons known only to you and your therapist, you prefer to believe in panspermia. In that case, life is also a widespread phenomenon, but will be somewhat reduced in structural variety.

LIFE OF A DIFFERENT SORT

Carbon-based life is what we know, and the medium in which it thrives and upon which it depends is water. But is it provincial to think that alien life will also be carbon-based and sloshing around on a wet planet? The fact that the overwhelming bulk of terrestrial protoplasm consists of just a handful of elements, the carbon, hydrogen, nitrogen, oxygen and phosphorous mentioned earlier, is not because these relatively light ingredients were the most plentiful on Earth. They weren't. But their chemical properties encouraged the formation of the complex molecules that underpin biology.

A reader steeped in science fiction will no doubt ask, "What about silicon life?" Well, what about it, indeed? Yes, it's possible that worlds too hot for liquid water have germinated silicon-based organisms, although silicon tends to make hard, crystalline structures like sand and quartz, which are not especially sympathetic to life. However, adventurous thinkers have imagined that long checkerboard chains of silicon and oxygen, with an occasional methyl (CH_3) group thrown in—silicones —might constitute a chemical backbone for life on torrid planets. Silicones are familiar in such incarnations as oils, waxes, rubbers, and breast implants. Lively, but not alive. Issac Asimov suggested replacing the hydrogen with fluorine, to make fluorosilicones that might serve as a basis for life on even hotter worlds. But let's state the obvious: silicon had its chance here, and didn't do much with it. There's more silicon than carbon on Earth. Sure, it may have been essential to life's early moments (in the form of clays), but we're not built of silicon. Carbon rules now.

So why make E.T. dependent on a chemistry that is difficult at best, and rubbery at worst? Carbon is known to be plentiful throughout the Galaxy. The cosmic abundance of hydrogen and oxygen assures us that watery worlds will be profuse in number. There are surely myriads of habitats in which carbon-based biochemistry could flourish. So while the details of E.T.'s metabolism will almost certainly be different from ours, a professional gambler would wager that his organic chemistry is the same.

INTELLIGENT LIFE

Once life gets started, there's a bit of a wait before it develops interest-ing mental capability and spawns thinking beings. On Earth, it took nearly four billion years for life to progress from the first organisms to *Homo sapiens*. But the time scale is less important than the probability. How inevitable was this progression from protist to professor? Could it be that life is common, but thinking, sentient life is rare? Could humans be the only reasoning creatures in the Galaxy, the only ones able to invent science and write poetry, simply because the evolution of intelligence is a chance event of very low probability?

This is an unsettled, and controversial, question. At first blush, it might appear that mankind, the crown of creation, marks the inevit-able endpoint of all that's gone before. It is a common belief for many that, given a life-bearing planet and enough time, intelligent beings will always arrive on the scene. But as evolutionary experts are fond of pointing out, the particular species that populate the Earth today were not inevitable at all. They are the result of chance, contingency, and circumstance. In his book *Wonderful Life*, paleontologist Stephen Jay Gould has emphasized the fact that if we could rewind life's tapes and start the drama of evolution anew, slight differences in climatic, tec-tonic, and even astronomical circumstance would result in a different assortment of flora and fauna today.

That point is indisputable. As an obvious example, if the dinosaurs and the majority of all other species had not been wiped out by the impact of a large comet or asteroid 65 million years ago, mammals would not have had their big chance to speciate, radiate, and produce the reader of this book. The Cretaceous dying was merely one fork in the multi-branched road that threads from the dim mists of the archeobacteria to the bright lights of the late twentieth century. On the billions of other Earth-like planets that we expect to exist in our Galaxy, the evolutionary roads will most assuredly have meandered along different routes.

Harvard zoologist Ernst Mayr, while optimistic about the frequent development of life on other worlds, believes that intelligent life is nearly miraculous, and therefore rare. His principal argument is that an enormously large number of life forms have developed on this planet, but only one is truly sentient. By 500 million years ago, virtually all the blueprints of modern life had been drawn, but, as Mayr notes, "Of the 40 or so original phyla of animals only one, that of the chordates, eventually gave rise to intelligent life." "In conflict with the thinking of those who see a straight line from the origin of life to intelligent man, . . . at each level of this pathway there were scores, if not hundreds, of branching points and separately evolving phyletic lines, with only a single one in each case forming the ancestral lineage that ultimately gave rise to Man." We are an accident, a highly unlikely accident, according to Mayr. If he's right, then we may be the smartest creatures in the Galaxy, no doubt a comfort to some readers.

However, the question is not whether other worlds have cooked up exact replicas of humans. That occurs only in low-budget science fiction. The question is whether intelligence of the sort that can understand and make use of science has evolved elsewhere. How often will Nature produce creatures with the characteristics that set us apart from "mere animals"—consciousness, cognition, and creativity?

The answer, despite Mayr's somber assessment, may be "often." Intelligence has sprung up on our planet in response to competitive pressure. When the first humans clambered out of the African savannas, they confronted an environment in which the number of species had reached an all-time high; twice as many as when the dinosaurs held sway. The world was a brutally competitive place. Brain power, like keen eyesight or sensitive hearing, conferred an advantage in this heightened battle for survival. Intelligence promotes adaptability. It may be a common solution to the challenges of a tough, terrestrial environment populated by sophisticated competitors. If life on other planets has similarly become diverse and competitive, it may also produce intelligent creatures.

This argument for E.T. is based on the concept of convergent evolution. In Nature, a common set of environmental circumstances will result in animals whose designs are analogous. For example, in the viscous undersea environment, we find a plethora of torpedo-shaped animals, for the obvious reason that streamlining increases speed. Both the ancient reptilian *ichthyosaur* and the modern mammalian dolphin are built to the torpedo plan, although their evolutionary origins are radically different. Oceanic creatures on distant planets would surely be streamlined, too. Austrian astronomers Mircea and Jorg Pfleiderer point out that the back legs of both cats and crickets are built with the same hinge and lever arrangements. This is another example of how Nature concocts analogous designs to solve a common problem. It is somewhat akin to the automobile industry. After a century of evolution, most of the cars on the road today are strikingly similar, with four wheels, piston engines which are usually situated up front, and so forth.

For those who are convinced that intelligence is a universally applicable solution to a frequently occurring environmental problem, such as intense interspecies competition, convergent evolution will see to it that we have some smart company in the cosmos. Dale Russell has played out this idea by considering what would have happened if the dinosaurs hadn't bought the farm. He has "evolved" a small Cretaceous dinosaur, *Stenonychosaurus,* to the present day. His rendition of this putative creature looks more like a human than like his *Stenonychosaurus* predecessor, and the suggestion is that smart dinosaur descendants might be running the planet now if astronomical catastrophe had not occurred 65 million years ago.

Ernst Mayr's counter-argument is that if convergent evolution were applicable to intelligence, then sentient beings would have appeared in more than one species, in the same way that eyes, hinged legs, and torpedo-shaped torsos have. Humans would not be the only intelligent creatures ever to walk the Earth. Astronomer Frank Drake rebuts this with the observation that someone has to be first. Is it really surprising that humans look around and note that we're the only smart kids on the block? Is the fact that we are the single intelligent species among the hundreds of millions that have sprung up on this planet remarkable, or merely a trivial consequence of being the first?

While the battle rages about whether life frequently spawns intelligence, we have to face the uncomfortable fact that argument alone is unlikely to settle the matter. What *could* answer this critically important question is the detection of an alien civilization, an enterprise that will be discussed in later chapters. Clearly, if intelligence is rare, then we are highly exceptional. But both modesty and the Principle of Mediocrity caution us about making such an assertion. And besides, it's an unsettling and unsatisfying thought that the crew of the Starship Enterprise might boldly go where no man has gone before, only to encounter dumb beasts.

Let us continue to suppose, as befits the premise of this book, that we truly share the universe with other intelligences. Would they be, as in *Star Trek,* human-looking folk clad in Greek tunics, spouting insipid philosophy in perfect English? Or would they have insect-like faces that only a plastic surgeon could love?

WHAT WILL THEY LOOK LIKE?

The only extraterrestrials that earthlings have yet seen are those appearing at the local cinema. But Hollywood aliens are like winter vegetables: there isn't much variety. Traditionally, most have been recognizably human, with heads, eyes, and multi-fingered hands. This preference for anthropomorphic aliens can be partially attributed to the practical necessity of designing a rubber suit for a human actor. However, modern screen aliens are frequently animated by computers rather than protoplasm. Hollywood's special effects teams can now endow extraterrestrial folk with convincing tentacles, steely teeth, and four quarts of mucous. Yet they often still look a little like the guy next door.

Does this reflect some sort of universal truth? Is there only one way to build an intelligent critter? Should we expect E.T. to look like us?

Let's consider what good engineering practice can tell us about an intelligent extraterrestrial's appearance. To begin with, we have asserted that any worthy E.T. must be able to hold up his side of the conversation. Consequently, we can assume a high degree of complexity, in a biological sense. Capability implies complexity. To prove this point, note that in the late 1970s, microprocessor chips combined thirty thousand transistors on a single, small slab of silicon. These chips could handle simple word processing and modestly-sized spread sheets. Two decades later, Pentium chips incorporated one hundred times as many transistors, and were able to dazzle their owners with snazzy graphics and multiple application windows (usually in the service of word processing or spread sheets). The universal link between complexity and ability ensures that E.T. will be multicellular.

This fairly obvious fact of intelligent life rules out the possibility that our alien pals will assume the shape of a malevolent mound of Jell-O, such as confronted Steve McQueen in *The Blob*. A single, large cell (as the carnivorous blob seemed to be) would surely be stupid. But of greater concern, the bulk of its protoplasmic innards would be situated far from its outer membrane. This would fatally hinder the exchange of gases and nutrients needed for survival, and the central parts of the blob would become gangrenous. Another liability for a single-celled sentient is the fact that any disease that succeeds in killing one cell will demolish the entire organism. Such critters are awfully vulnerable.

Clearly, enormous, blob-like cells are non-starters. And at the other end of the scale, there's a limit on how *small* a cell can be, too. *Mycoplasma* bacteria, microscopic organisms that measure about 10 millionths of an inch across, are typical of Nature's best miniaturization efforts here on Earth. A substantially smaller cell would have a volume too puny to support the necessary chemistry of existence, a fact that will also apply to aliens. Consequently, since we anticipate that the complexity required of a thinking being will involve at least billions of cells, there's no chance that E.T. will fit on the head of a pin.

Another reason to expect that intelligent extraterrestrials will be large enough to trip over derives from simple geometric scaling. If you take any animal and somehow shrink it to half its normal dimensions (that is, reduce its size by a factor of two), you'll find that its surface area—the amount of skin needed to cover it—has decreased by the square of the size change, or a factor of four. In other words, the shrunken animal ends up with a higher epidermis-to-endoplasm, or skin-to-innards ratio. Like an automobile radiator, it will lose body heat quickly. For this reason humming birds eat constantly and copiously, whereas elephants don't. This sets a limit on how small a warm-blooded animal can be. If they're less than a certain size, they won't be able to eat fast enough to avoid a fatal cooling off.

This is relevant because E.T. is probably warm-blooded. Thermostatically-controlled creatures are the high-performance members of the animal kingdom. They are both more active than their cooler brethren and better able to adjust to climate variations. So if you believe that E.T. prefers the faster lifestyle and greater adaptability that comes with being warm-blooded, then it's unlikely he'll be much smaller than a humming bird. Indeed, you should expect him to be considerably larger. If he's bird-sized, he'll probably be bird-brained, and not terribly bright. He'll also be spending all his time the way that humming birds and mice spend theirs: foraging for their next meal.

The scaling laws set a limit to E.T.'s *maximum* size too, since they govern his agility. Imagine enlarging a creature to twice its normal dimensions. Strength, which depends on muscle cross-section, will grow by the square of the size increase, or a factor of four. This sounds pretty good until you consider that the scaled-up creature's weight, which goes as the cube of the size, has increased by a factor of eight. The animal now has a power-to-weight ratio that's half of what it was, thus reducing physical ability. This simple scaling explains why an ant can carry many times its own weight, while an elephant can't. If you drop an ant from ten times its height, it will hardly notice (mind you, ants barely have a brain, and don't seem to notice much of anything). Drop an elephant from ten times its height, and you will be faced with an unsettling clean-up. Very large animals are impressive, but not often for their agility. They depend on massive appendages to stand up and move around. (For this reason, scaled-up ants, spiders, and grasshoppers, such as occasionally appear in Hollywood creature features, are impossible. The relatively spindly legs of these giant insects wouldn't allow them to get off their bellies.)

Elephants, whose legs are chunky enough to be turned into umbrella stands, are currently the largest creatures walking the Earth. The only living entities with greater heft hang out underwater (whales) or underground (fungi). Either environment provides help in supporting Brobdignagian bulk. But for obvious reasons, such sub-surface giants aren't likely to develop astronomy or make sophisticated tools. These arguments suggest that if E.T. inhabits an Earth-like planet, it's most probable that, size-wise, he'll be somewhere between ten pounds and ten tons. This might be scaled up or down a bit depending on his planet's gravitational tug, of course.

In the discussion so far, we've made the implicit assumption that E.T. is an animal. Animals have mobility, and we expect that ambulatory life forms will be the first to develop intellect. A dog, which has to find and catch its next meal, requires more brain power than a dogwood. So our sentient extraterrestrial will have some method to get up and go. On Earth, most locomotion for larger animals is afforded by appendages, either legs, wings, or fins.

A detectable, intelligent E.T. with fins is unlikely. Underwater creatures might never put together the technology required to visit or signal distant worlds. (One impediment is the fact that radio waves don't penetrate sea water). Marine environments have also been characterized as "too easy," and not prone to produce smart creatures in a hurry. In the ocean, movement is relatively simple, the temperature changes only slowly, and the weather's always the same. Most probably it will be land animals that first develop substantial brain power—intelligence will be the evolutionary response to the climatic and topographic rigors of the high and dry. In support of this idea, note that dolphins, considered to be among the brightest of briny folk, didn't evolve underwater. They were land-dwelling mammals that returned to the sea. The stark facts are that most marine life is dumb, and even the celebrated cetaceans haven't produced either science or great literature. So it seems safe to say that space-faring extraterrestrials who live underwater, such as featured in the movie *The Abyss,* are all wet.

Might E.T. have wings? Flying aliens are certainly a possibility, although the heaviest airborne animals on Earth weigh in at twenty pounds or so, and animals of this heft seem too small to have a great deal of intelligence. On the other hand, planets with thick atmospheres, lower gravity, or especially oxygen-rich air to turbocharge metabolism, might conceivably spawn some soaring intelligence.

Having considered the air and the sea, we return to everyone's favorite extraterrestrial habitat: dry land. For locomotion on terra firma, legs and their derivatives, arms, are the equipment of choice. No animals have wheels, despite their efficiency. Wheels function well only on prepared geography, either rails or roads. (Snakes manage to get around without either appendages or wheels, but snakes don't make tools.) The number of appendages, normally four for the bigger animals on Earth, should be large enough to permit both movement and manipulation (thus, more than two), and small enough not to burden E.T.'s brain with an enormous processing load (less than, say, a dozen, assuming they have individually movable digits). Insects sport six appendages, and dominate the species count on our planet. So six appendages is a number Nature surely finds agreeable. The fact that humans, hippos, and hyenas all have four extremities is the accidental consequence of the evolution of backboned animals from a four-finned ancestral fish, *Eusthenopteron*. Had *Homo sapiens* been blessed with six appendages, we would be more adept at piano duets and handball.

We can also expect an intelligent alien to have eyes, useful for gathering information about the local environment. Indeed, eyes are so handy that Nature independently developed them in many dozens of species on Earth. They seem to be an inevitable product of evolution in a world flooded with light. In the previous chapter we presented reasons why it's most likely that advanced life will be found on planets orbiting Sun-like stars. In such cases, the spectral quality of the light illuminating these worlds will be similar to what we experience on Earth. If the atmosphere lets at least some of this sunshine in, we can anticipate that E.T. will have an eyeball or two. Actually, two is much better than one, as a pair will offer depth-perception, always an advantage when it comes to catching dinner. The benefit of having more than two needs to be weighed against the cost of the additional cerebral processing. None of the higher animals on Earth have opted for more than two eyes.

Hearing is another universally useful talent because it permits communication in environments, such as jungles, where sight-lines are short. It also allows you to warn someone whose back is turned. Smell, although not overly developed in humans, is widely used by other animals, and can convey highly specific information. The disadvantage of relying on scent alone is that odors are wafted to you via diffusion in the air, and are therefore hard to pinpoint. It's easy to know that there's a dead skunk somewhere along the highway. It's less easy to know exactly when your car passes it on a dark night.

The gist of this discussion is that all the human senses also make sense for E.T. And it's only good engineering practice to put the sensory organs—eyes, ears, nose, or their extraterrestrial equivalents—up high where they can get a better "view," and to situate them close to the brain to minimize reaction time. In other words, E.T. will have a head, and his brain will be in it.

This discussion may seem to vindicate Hollywood's uniformly humanoid aliens. It is, however, no more than an argument in support of convergent evolution. Since E.T. is likely to depend on carbon-based biochemistry, he'll be ensconced on an Earth-like planet. This similar environment will lead to analogous adaptations. And there's no disputing the fact that *Homo sapiens* is a decent enough design. More than three billion years spent budding and pruning the evolutionary tree is bound to produce good designs. But the human form is certainly not the only one that would work. A visit to the local zoo will confront you with hundreds of complex animals, creatures that are quite literally our genetic cousins. And yet practically none of them look much like us. Despite the fact that we share both a planet and a lot of DNA, only a minuscule few of these fellow earthlings would be mistaken for a human ten feet away on a foggy night.

And yet, a lot of movie aliens *do* look like us. Their appearance has little to do with universal laws of sentient design. It is a projection of our fears. According to Hollywood, space is dominated by two types of inhabitants: the good aliens who come to Earth to enlighten or amuse, and the bad ones who want to wreck the neighborhood, colonize the neighborhood, or simply chow down on the neighbors. As befits all cinema protagonists, the white hats among the extraterrestrials are good looking, and the black hats are repulsive. In this case, good looking means anthropomorphic. Mr. Spock, the famous Vulcan from *Star Trek*, is indistinguishable from a human, despite his quirky ears and *outré* page-boy haircut. Television alien *Alf* shares with *E.T.* a small stature, wrinkled skin, big eyes, and a short nose—the characteristics of human babies. Indeed, the features of infant chimps, puppies, and newborn humans are all similar: a big head, small body, short legs, a round face, big eyes, small nose and mouth, a lack of hair, and an expansive forehead. These common characteristics serve to elicit a nurturing response from the parents. If a film alien fits this description, he's obviously cute, harmless, and here to help.

Deep-space villains, on the other hand, often have features that are insectoid or reptilian. In *The Arrival* and *Independence Day,* the basically pseudo-human aliens are outfitted with disgusting tentacles. Curiously, tentacles are probably a bad choice for an advanced extraterrestrial, as muscle organs without bones are less mechanically efficient and less adept for grasping than hands. But movie audiences have a

The baby-faced, benign alien from *E.T. The Extraterrestrial.* [Academy of Motion Picture Arts and Sciences]

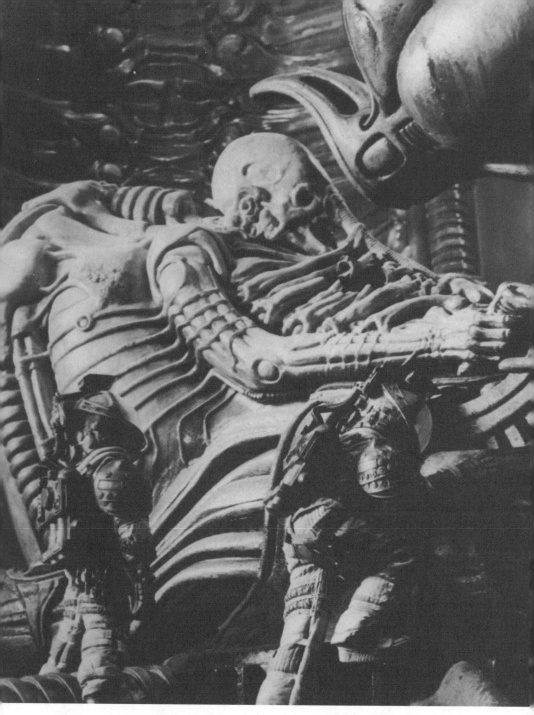

Earthlings examine the decayed remains of an extraterrestrial in the movie *Alien*. This particular creature was presumably an earlier victim of the film's toothy bad guys. Despite sporting a nose like a vacuum cleaner hose, this victim looks remarkably anthropomorphic. The skeletal remains of his arms, fingers, rib cage and cranium are patterned after human design, an unlikely circumstance for any real aliens.

common repulsion to snake-like life, perhaps because such creatures once slithered into the trees to prey upon our ancestors. We have a similar aversion to insects, which bring pain, disease, and constant annoyance.[4] Bad movie aliens frequently ooze slime and mucous, ciphers for sepsis and hungry carnivores. All in all, there is about as much subtlety and realism in Hollywood's cast of characters from space as in its depictions of the Old West. Physiognomy recapitulates motive.

ABDUCTING ALIENS

In addition to Hollywood's heavies and heroes, popular culture embraces a third class of aliens. These are the supposed occupants of the UFOs that buzz the countryside. Clearly, these elusive visitors are not interested in laying waste our cities, for otherwise they would have already done so. Their intent is more enigmatic. Sometimes they seem keen to give tours of their saucers to random human guests, and occasionally they poke and prod these on-board visitors or offend their sensibilities by extracting eggs or sperm.

While the popular description of these putative visitors used to be "little green men," the color *du jour* is now multiculturally-correct gray (perhaps a reflection of their status somewhere between the good and the bad aliens of Hollywood). Once again, their beauty, or lack thereof, is in the eyes of their human beholders. Emory University biologist Lori Marino believes that the "grays" are a projection of what we think humans will become with a bit more evolution. "They are hairless, and after all many of us face hair loss. They have a tiny nose and a small, toothless mouth, because our own olfactory sense is diminishing and we are losing dentition."

The grays have big eyes, since in the age of television, vision is primary. Their big heads, covering big brains, sit atop small bodies. This isn't

---

4. It is interesting to note at least one occasion on which Hollywood diverged from this association of danger with reptiles and insects. A 1972 film, *Night of the Lepus*, told the story of mutant rabbits, dozens of feet high, rampaging through a western town. Forsaking their normal, vegetarian eating habits, the bulked-up bunnies turned to humans for their next meal. The film was ludicrous, rather than terrifying, because even in giant format, bunnies are still bunnies.

merely the baby build described earlier. These guys are mentally strong, but physically weak. They are geared to our perceived future, when brains, not brawn, is the winning characteristic. Grays are the type of critter that might take your job.

"These aliens are what we expect our descendants to be like," Marino points out. "But they are also very cool, without emotion. And an emotionless, loveless future may be what scares us the most. The grays have dark eyes—insect eyes—black pools that betray no feelings."

Their saucers may be hi-tech, but these smooth-skinned visitors from other worlds are emotionally challenged. They are gray-skinned zombies. In fact, this is probably wishful thinking, because emotions are no less expected in the makeup of an advanced creature than hair, ears, or eyes. There is a tendency (most pronounced among men) to view emotions as a weakness, a bungle in our design. If we could only be completely logical, like *Star Trek's* Mr. Spock, we'd be better for it, in this view. But in fact, passion has a purpose. Emotions are useful for survival, motivating us to reproduce, to care for our mates and our young, and inciting us to destroy predators and neighbors intent on muscling their way into our territory. The absence of alien emotion isn't realistic, but it makes these creatures from afar both scary and reassuringly inferior.

Hollywood knows what extraterrestrials will look like, as do the UFO buffs. But when science is pressed to picture our cosmic neighbors, the result can be but shadowy hints and suggestions. Extraterrestrials will, as we've seen, very likely share much of our biochemistry. They will have evolved in an environment that can support that chemistry, on a planet not too dissimilar to our own. The aliens will be built of protein-rich liquids, encapsulated in membranes that define cells. They may be larger or smaller than we are, but not by more than a factor of ten or twenty. Their complexity and range of senses will likely be at least the equal of ours, and their sense organs and brains will be centralized in a head. In order to better see what's coming and what's available to eat, that head will be held high. E.T. will have appendages for locomotion and manipulation, but how many is uncertain. More than two and less than a dozen is a reasonable range. The extraterrestrials will have eyes, ears, and emotions. Their skeletons will be on the inside, not on the outside like an insect. An internal skeleton allows growth without inconvenient (and dangerous) periods of molting.

All this we can expect on the basis of good engineering practice, and in fact, these arguments are more than sufficient to rule out many of the more extreme extraterrestrials of science fiction. But what we cannot expect is that the aliens will look like us. The diversity of life on our own planet is adequate warning. The cut of E.T.'s jib may be as exotic as the creatures we find caged and penned at the zoo. We can no better predict the exact style of his physiognomy than we could have predicted the appearance of the tube worms.

ALIEN REPRODUCTION

As long as we are prying into the matter of E.T.'s looks, we might as well consider his sex life, too. This is of obvious interest to the public, given the widespread belief that UFOs are occasionally ferrying extraterrestrials to our planet to abduct us for breeding purposes. In Ridley Scott's dark film *Alien,* this idea was turned around. A spaceship with a (mostly) human crew was lured to a punk planet where they found the eggs of an extraterrestrial species. Tampering with one of the eggs led to the unwilling ingestion of an alien embryo. This was first perceived to be but a minor annoyance, but the embryo grew and finally hatched Cesarean-style from its human host. Once loose in the spacecraft, the *enfant terrible* strived to satisfy its nutritional needs by dining on the crew.

So here we have the circumstance of humans traveling to other worlds to be used for breeding purposes. Either way—UFOs or planetary eggs—it seems that our interaction with aliens is often to help them reproduce. Is there any basis for this? Does sharing the universe also entail sharing our beds?

Of course breeding with aliens, whatever the aesthetic merits, won't work, a fact that will be obvious to any alien worth his carbon. While their biochemistry may be similar to ours, the details of their inheritance mechanism (in our case, DNA and RNA) will most assuredly be different. Recall from high school biology the definition of a species: it is the ability to successfully interbreed. King Kong may abduct Fay Wray for salacious purposes, but he won't be successful (even aside from matters of consent or fit), despite the fact that as primates, King and Fay have more than 99% of their genes in common. The aliens will have 0% in common with either.

One could argue that the alien in *Alien* wasn't really interested in shuffling its genes with ours. It merely wished to use our bodies as incubators. This film alien clearly took a cue from the earthly ichneumon fly, an insect that shamelessly (and with fatal effect) lays its eggs in a caterpillar. Such parasitic procreation is appealing to Hollywood, since it is repulsive to humans. Even before birth, this alien is clearly a bad guy. But for the extraterrestrial, such a complicated reproductive strategy is risky. It has to hope that its eggs are found by interstellar travelers (and are not tasty enough to be turned into omelets). If spacefarers of the right sort don't show up often enough, the eggs could go stale, and another litter of toothy aliens will be lost.

Also noteworthy was the low yield. A rocket full of humans finds a room full of eggs, and the net result is the production of one—count 'em, one—adult alien. In addition, the intelligent sort of incubators this alien seems to favor might some day post a notice on the galactic internet, warning potential hosts to avoid strange clutches of eggs (something caterpillars are unlikely to do). Bottom line? Depending on a (mildly intelligent) host from afar is no way to have kids. Imagine if humans were dependent on itinerant chimpanzees for reproduction.

One aspect of *Alien's* chancy spawning cycle that the film mercifully omits is the production of the eggs in the first place. This might involve sex. If so, the alien would eventually confront the problem of locating a mate somewhere else in the Galaxy, a dating challenge that most introduction services would find daunting.

It's also possible that alien reproduction doesn't involve sex-as-we-know-it at all. On Earth, the majority of complex organisms are, indeed, bisexual—separated into two sexes. By melding genes from a pair of individuals, progeny are spawned with a combination of characteristics proven to have survival value in their parents. (This is generally a better deal than introducing new characteristics via DNA replication errors or by having an occasional cosmic ray smash into a gene.) Sexual reproduction, whatever its other attractions, speeds evolution. But some creatures, notably aphids, worms, bees and wasps, use different strategies. Aphids usually reproduce asexually, producing only females. Once a year they binge, and enjoy a little copulation to shake up the gene pool. Some worms are hermaphroditic, and combine both sexes in one individual (although you'll be relieved to know that they don't mate with themselves). Bees and wasps produce females with two sets of DNA, but males with only one. This means that females share three-quarters of their genetic inheritance with other females (but only one-fourth with the males). Female worker bees can be expected to toil hardest for the survival of their sisters, the working class heroines of the swarm. Consequently, this genetic strategy may strengthen the bees' rigid social structure. Since we often imagine the aliens as being members of highly organized societies, maybe they reproduce more like the bees than the birds.

Possibly E.T. has reached such a state of perfection that offspring with novel characteristics might be a disappointment to their parents and a threat to their species. For such a perfected race, asexual reproduction would be the way to go. Fission *á la* bacteria would be difficult and probably painful for a complicated organism, but budding is an alternative.

Such speculations, driven by the wide variety of reproductive schemes on Earth, should cause us to beware of the facile assumption that the aliens will be of two sexes. It is highly unlikely that they would be eager to get their hands, or any other body parts, on human gametes. The aliens might be interested in us for breeding *experiments*, but not for breeding *purposes*.

What, then, might they want from us?

Alien Abilities and Behavior

In the 1996 blockbuster film *Independence Day,* the American president finally manages to have a short but pithy conversation with one of the aliens whose machines are wreaking havoc and destruction on our planet. "What is it you want us to do?" asks the president in measured tones. The alien responds pointedly: "Die."

The message was brutal, but you have to give the extraterrestrials a lot of credit for knowing their mind, and being straight up about telling us. You also have to marvel at the fact that such a conversation could take place at all. This is not so much a question of language (the alien uses a human intermediary as a translator), but rather of mind-set. The alien's mental capabilities are obviously not so different from our own.

This is a common circumstance in movies. In the 1985 film *Enemy Mine,* earthling Dennis Quaid finds himself stranded on the less-than-lovely planet Fyrine IV with only a reptilian alien for company. By the film's midway point, the earthling and the extraterrestrial (played by Louis Gosset, Jr.) have joined forces in the interests of survival. The alien soon becomes the best buddy Quaid ever had. Not only do the two species understand one another, but they share such traits as anger, humor, and a knack for sarcastic insult.

Hollywood aliens inevitably think like humans, despite their evil intentions and scaly complexions. Part of this resemblance is due to the requirements of dramatic structure. If your antagonist is unfathomable, then his actions appear to be random. Such an enemy cannot be either evil or outsmarted, for he has no comprehensible will. Even the great white shark in *Jaws* had a clearly understandable game plan: to gnash and gnaw his way through as much of the film's supporting cast as possible. He wasn't just an underwater eating machine, he was malevolent. That set the audience up for a collective sigh of satisfaction when the great white was turned into fish meal at the movie's explosive climax.

Real sharks, of course, are less deliberate, and also less decipherable. Humans have a hard time imagining how fishes or birds think about the world. The only concept of "thinking" we can muster is our own. Consequently, we assume that our own perceptions and style of comprehension will be shared by E.T. Needless to say, they may not be.

E . T . ' S   C U L T U R A L   L E V E L

If we ever encounter or communicate with extraterrestrials, one thing is overwhelmingly probable: the aliens won't be at our level of development. Neither their technical abilities nor their IQs will be comparable to ours.

Why do we say this? Simply because if we detect the aliens, the very fact that we've done so will be a consequence of their high degree of sophistication. To begin with, it's obvious that any extraterrestrials we uncover can't be less sophisticated than humans. We won't find backward, primitive aliens on our doorstep or hear them on the radio, simply because we will only encounter E.T. if he makes contact with us, either literally or figuratively. And that requires substantial expertise.

For example, it's possible (though unlikely) that extraterrestrials might physically enter our solar system and pay a personal visit. Perhaps they will land in the backyard and ask to be taken to the Dalai Lama or some other leader. That would involve a feat of interstellar transport far outstripping our own abilities. Any aliens capable of rocketing from one star system to another will obviously hail from a very advanced society. If E.T. pays a house call, you can be sure that, at least on a technological level, he is very different from you and me.

Another prospect is that we may detect E.T. at a distance, for example by eavesdropping on his radio traffic. That, too, would require that he be scientifically adept. To give us a chance of tuning in on his signals, E.T. would have to construct powerful transmitters and affix them to large, steerable antennas. So clearly, even if we're expecting communication rather than contact, the aliens will need to be at least as technically sophisticated as twentieth-century humans.

In fact, they will be a lot *more* sophisticated. The reason for this can be understood by a simple analogy. Imagine that you have just learned the game of chess. You memorize the moves, play a few games against your younger brother, and decide that you really like this kingly pastime. So you find a chess club in town, and join up. With the club's roster in hand, you randomly pick one of the members to challenge for your first, friendly game. The chances are high that your initial face-off will be against someone who has played chess for years; someone who has advanced far beyond your level of competency. You are the new guy in the club, and most of the other members have a lot more experience.

When it comes to communicating with E.T., we are like the chess neophyte. Humans have had high-powered radio transmitters and sensitive receivers for only a few decades. So we're still a new member in the club of radio technology. The first other clubman we encounter may not even remember which of his ancestors *invented* radio. Any extraterrestrial with radio capability will likely have had it far longer than we.

In addition, if we do pick up E.T. on our radio telescopes, that very fact will virtually guarantee his advanced cultural standing. Imagine, as example, that every 10,000 years a planet somewhere in the Galaxy cooks up a technologically sophisticated civilization. This example postulates a fairly optimistic birth rate, incidentally, for it implies that the Galaxy has witnessed the appearance of a million such civilizations in the course of its long history. Now, you can be certain that each of those technological societies will invent the atomic bomb at about the same time they invent radio. After all, the two devices appeared on Earth only a generation apart. If you are the pessimistic sort, you might expect that many of these newly arrived technological societies will blow themselves up within a century or two after first going on the air with their high-powered radios. If this regrettable situation is the norm, then societies will only broadcast for a few hundred years. The chances that *any* of the million civilizations that have sprung up in the Galaxy is on the air now would be only about 1 in 100. We won't hear a thing, and any receivers we use in an attempt to tune in E.T. will be greeted with nothing but static.

On the other hand, suppose you take a more optimistic view. Imagine that technological societies manage to survive and stay on the air for ten thousand years or more. Then clearly we might have a reasonable chance of hearing one, since at least a few simultaneously broadcasting societies will be hanging out in the Galaxy at any given time.

The numbers we have cited in this argument are only an example. After all, no one really knows how often technically sophisticated societies arise. But the argument itself leads to a conclusion that is not critically dependent on the precision of our estimate. It boils down to this: If we succeed in overhearing an alien broadcast, the odds are great that the beings on the transmitting end are members of an old civilization. For them, radio will be a technology that their ancestors devised tens of thousands, or possibly many millions of years ago. They will be as much in advance of us scientifically as we are beyond the Neanderthals, and maybe a great deal more. In addition, if this difference in maturity is large, it may be that they have outstripped humans in more than just technology. Their intellect may also be far suppler than our own.

E . T . ' S   I . Q .

Is it possible that E.T. surpasses us in raw I.Q. as much as we surpass caterpillars and canaries? The human experience suggests that he could. After all, brains have developed quickly for humans. *Australopithecus africanus,* one of our humble predecessors, had only a pound of gray matter. In the two million years since this pint-sized pundit roamed the savannas, the human brain has trebled in size, and now houses tens of trillions of neurons. So why not bring on more? Unfortunately, the dimensions of the human birth canal limit how large a baby's head can be. Big-headed babies can't be born. And that limits the number of neurons, as evolution on Earth has so far failed to miniaturize these building blocks of the brain (they are the same size for all animals). If Nature is making any attempts to produce significantly smarter humans, her efforts may have stalled, at least for the moment.

But that needn't be the case for E.T. If he can somehow reduce the component size of his thinking machinery, for example with smaller neurons, then he could pack more brain into his head. And of course, his head could be bigger to begin with. Either of these factors could send E.T.'s IQ off the charts, as a small change in neural number makes a big difference in performance. While our cerebral volume is only three times that of *Australopithecus,* our IQs are far more than triple his. When it comes to thinking, what matters is the number of neural *interconnections.* The number of interconnections rises geometrically with the neuron count. As Iosef Shklovskii and Carl Sagan noted many years ago, there is probably no limit to the extraterrestrials' intelligence. So while it's likely that other technological societies may have been started by creatures with human IQs, any that we overhear or encounter will have had a significant amount of additional time in which to dramatically improve their intellectual lot.

This disconcerting thought is one that Hollywood is unwilling and unable to address. Cinematic aliens are always at about our level of savvy, although they frequently outpace us in technology. (Not *all* technology, however. The hostile extraterrestrials of *Independence Day,* despite having built the mother of all mother ships, eventually succumb to 20th century computer viruses and dogfight tactics.) Only a modest amount of reflection should convince you that the aliens of the movies are of necessity at our level, for otherwise the films would be, like *Bambi Meets Godzilla,* short and distressing. It is good dramatic practice to have a worthy antagonist. It is bad dramatic practice to have one that's invincible. Alas, the real aliens are much more likely to fall into the latter category than the former.

Of course, some extraterrestrials will be more advanced than others. The differences will not be minor. On Earth, all human societies are similar in their level of development. Yes, the *National Geographic* can make a decent profit chronicling the customs of primitive tribes in some antipodean outback, but even these stone-age societies differ from their first-world opposite numbers by only a few millennia at most. *Homo sapiens* is a mere 100 thousand years old. All mankind left the starting gate at the same time. As different as human societies may seem, they're still running very much in a pack.

Not so for extraterrestrial societies. The Galaxy has been around for 10 billion years or more. Presumably, sentient civilizations have emerged during most of that long history. Needless to say, it's extraordinarily unlikely that two random worlds, separated by hundreds or thousands of light-years, will develop in parallel. No two societies will have left the starting gate at the same time. The disparity will be such that there is little chance that aliens from two societies anywhere in the Galaxy will be culturally close enough to really "get along." This is something to ponder as you watch the famous cantina scene in *Star Wars.* In this futuristic pub, a politically correct, multi-hued assemblage of extraterrestrials (all conveniently upright in posture, if not demeanor) share a brew and engage in some back-slapping camaraderie in the grungy port city of Mos Eisely. Does this make sense, given the overwhelmingly likely situation that galactic civilizations differ in their level of evolutionary development by thousands or millions of years? Would you share drinks with a trilobite, an ourang-outang, or a saber-toothed tiger? Or would you just arrange to have a few specimens stuffed and carted off to the local museum?

The fact that E.T.'s intellectual plane will be beyond that of humans may make him difficult or impossible to understand. Even aside from the indelicate matter of IQ, the enormous culture gap could cause alien communications to appear mysterious or magical. Consider for example Cro-Magnon man, who sported biological equipment quite similar to our own. Despite the kinship, Cro-Magnon man would have a hard time understanding the purpose of a computer operating system or the meaning of a television signal. Two centuries ago, the scientific societies of Western Europe began to regularly interact with one another, thereby stimulating a faster research pace. The societies were able to do this because the countries involved were all at about the same level. They also spoke comprehensible languages. Neither condition will obtain in the case of communication with aliens. Human intelligence has been so shaped by the peculiarities of our Earthly environment and evolution that we might be simply unable to connect with non-humanoid intelligence. And, needless to say, the aliens aren't going to speak languages understandable to us, nor become instantly fluent in English. In fact, they might be physiologically incapable of speaking *any* Earthly languages, in the same way that a giant squid is.

Aside from problems in communication due to lack of vocal chords or compatible ways of thinking, there's also the matter of common cultural ground. E.T. might regale us with music or poetry, but there's little chance that we'll either recognize or appreciate it.[5] The alien's literature and even medicine may have little congruence with, or relevance to, our own.

However, sci-fi stories have conditioned us to expect at least some points of cultural overlap with E.T., especially in science and technology. As noted, we demand that the extraterrestrials be accomplished in such fields. These minimum requirements spring not from our desire to promote science literacy in the Galaxy, but simply because only technically deft aliens will make their presence known. In the last chapter, we argued that intelligence will frequently spring up on planets where species compete, thanks to its tremendous capacity for adaptation. But does intelligence inevitably lead to science, particularly science like our own?

Once again, researchers try to gain insight on a universal question by holding a mirror up to themselves. They turn to our own history, and what they find is that science is rare for human societies. While every earthly civilization worthy of the term has elaborate social structures, organized agriculture, and urban centers, they don't all manage to

5. Note that this hasn't kept us from occasionally launching some of our own music into space. The Voyager probes, sent to reconnoiter the outer solar system in the late 1970s, carried a small record bearing such cosmopolitan musical offerings as Bach's *Brandenburg Concerto* and Chuck Berry's *Johnny B. Goode*. These were for the edification, or perhaps puzzlement, of any extraterrestrials who might retrieve these spacecraft in the distant future.

cobble together an analytic description of the physical world. In Europe, the millennium-long preoccupation with theology during the Middle Ages promoted very little scientific development, despite the widespread practical knowledge of many artisans. Anthropologist Kent Flannery notes that the Mayans had civil-servant astronomers and boasted a better calendar than the Spanish conquistadors who eventually laid waste their state. But the Mayans didn't develop metal tools or even the wheel. Philosopher Nicholas Rescher finds it easy to imagine advanced alien societies that are possibly skilled in engineering, but lacking in substantial science. As Rescher says, "A lot of know-how can be built up without much know-why."

This skepticism regarding the emergence of science is reminiscent of the argument against the evolution of intelligence. "You don't need it to get by, so it might not occur very often." But the counter-argument is also similar: an understanding of nature can rapidly expand your range of actions. Intelligence has survival value, and so too does science. Less than four centuries after Copernicus espoused a new way of looking at the heavens, we see modern science's dramatic effects in every corner of the planet. It has revolutionary consequences, not the least of which is the rapid extension and proliferation of our species. And while the majority of Earth's civilizations, including such long-lived empires as Egypt and China, failed to develop very much science, it does not seem to be a fluke, an improbable activity limited to Renaissance Europe. The Greeks, at least, had been there before. Human intelligence took four billion years of evolution, and has appeared once. Still, we think it likely that comparable intelligence will sooner or later emerge on many inhabited planets. Science has taken intelligent creatures only 100 thousand years to concoct, and has appeared more than once. So it seems that if intelligence is probable, science is inevitable.

E.T.'S ATTITUDE

Aside from matters of a cultural and cerebral nature, can we say anything about the extraterrestrials' world view? We've noted that any aliens that we either detect or run across will be members of an old, technologically accomplished society. Given this elevated cultural status, mightn't we expect E.T.'s morals and motivations to be on a similarly lofty plane?

We might, although Hollywood usually doesn't. Film aliens are generally of small moral caliber. The malevolent ones are single-minded predators, with only destruction or lunch on their minds. Their ethics are no more complex than those of a piranha. They don't feel much need for sociable behavior even among their own kind. Hollywood extraterrestrials seldom interact with their planetary brethren, opting instead to follow one another around like mindless robots with strict orders.

The friendly cinematic aliens, on the other hand, share human morality so completely that they degenerate into funny-looking pals who happily participate in our activities. Little *E.T.* will pull a beer from the fridge, sit on the couch and watch daytime television. The furry Chewbacca and the logical Mr. Spock willingly take on the second-in-command jobs in our spacecraft. We readily accept these alien comrades, although you would probably regard it bizarre in the extreme if NASA suddenly opened its astronaut program to non-human species here on Earth, let alone to something reared on the eighth rock from Alpha Centauri. Could we rely on such extraterrestrial crew members in a crunch? Would they give their all to save human society? Can we expect that they have the same ideas about winning, about power, friendship, or courage?

What we call moral behavior, or altruism, clearly has survival value in human society, and consequently the real aliens will likely practice it as well, at least with one another. Of course, it's asking a lot of E.T. to ask him to extend his altruism to non-related species from another solar system. But perhaps such altruism could be motivated by religious belief. Religion, which institutionalizes morality, promotes stability in human societies. It may do the same for the aliens. In the 18th century, as Captain James Cook explored the Pacific discovering island after island, his log invariably contained descriptions of the local religious practices because such practices were always to be found. Religion is universal among human societies. Whether the same holds true for the extraterrestrials depends on whether religious belief is a fundamentally useful survival tool, or merely an artifact of some other human trait. For example, music, like religion, is ubiquitous among earthly societies. But music doesn't really promote survival; it's a side benefit of the way in which our brains are organized for other activities that *do* promote survival. Music may not be among the pastimes of thinking beings organized along different lines. In the same way, it's difficult to know whether the extraterrestrials will have their own theology.

Of course, if they do, there's little chance that the specifics of E.T.'s faith will mimic our own, any more than his appearance will resemble ours. Not so long ago, this self-evident statement was considered less so. In 1853, British philosopher William Whewell, who penned *Of the Plurality of Worlds: An Essay*, was concerned that if the universe is filled with intelligent societies, then Jesus would be compelled to undertake an endless voyage of salvation from one planet to the next. Whewell realized that this literal extension of Christianity to other worlds might considerably dilute mankind's special relationship to God. Earth would be just one more station on a non-stop itinerary to save souls. However, the interest in, and concern with, logistics problems of the sort that bothered Whewell have dissipated in the twentieth century. An informal polling of modern theologians by the Israeli sociologist Michael Ashkenazi has shown that today's clergy members are far less insistent that alien religions should conform to earthly ones.

Although the extraterrestrials would probably be altruistic with one another, and even religious, that possibility alone says very little about whether they would be "moral" with us. Altruism has a biological basis, as noted. A selfless act may prompt reciprocity at a later date. One good deed encourages another, and both the individual and the species benefit. However, when it comes to interactions between extraterrestrials and humans, the aliens will have little biological reason to be altruistic, only intellectual ones. After all, consider how we treat animals. Some are our pets. Others we grind up and lace with ketchup. Human attitudes about the treatment of our local "aliens," namely animals, are highly divergent. We treat many of these other species badly (from their point of view). Why shouldn't the aliens behave the same way towards *their* aliens?

This argument might be faulted on the grounds that the animals we abuse are, after all, not terribly intellectual. Presumably, any E.T. in contact with us will recognize our cognitive horsepower, and grant us a little respect. We're all sentient beings, after all, and surely we can sympathize with one another?

Not necessarily. Altruistic behavior towards fellow beings who are outside the immediate group is not so obviously advantageous, and doesn't always occur. The diplomat Ichiro Kawasaki has noted that traditional Japanese society is subject to rigid rules of etiquette and behavior. How, then, to reconcile this institutionalized civility with the brutalities perpetrated by Japanese soldiers while occupying other Asian countries in the Second World War? Kawasaki suggests that "once outside the confines of his home or family, a Japanese is at last 'liberated' from all these restraints and starts behaving like a different person." The rules of behavior at home don't necessarily apply abroad. The indigenous inhabitants of the Americas and the islands of the Pacific all suffered grievously when European explorers made contact. This was true even during the voyages of Cook, who was under specific instruction not to inflict damage on the natives. The aliens aren't guaranteed to treat us gently simply because we're conscious and cogitating.

Curiously, the distress that befell the South Sea islanders was not really Cook's fault, and in his story lies a lesson that might be applicable to any future contact with extraterrestrials. The famous British captain had insisted upon honest dealings between his crew and the natives, and he absolutely forbade violence. He also refused to let syphilis-infected sailors ashore. However, this enlightened effort to avert later disaster was frequently thwarted. The native women would swim out to the ship. They did this because the foreign civilization that had appeared on their coasts was seen to be technologically superior. After all, they arrived in large sailing ships and were outfitted with cannon. The native women assumed that these technically proficient visitors were superior in every other way, as well. They might be good breeding stock.

The presumption that alien technological prowess goes hand in hand with a superior culture or sophisticated morals is one that we might make as naturally as did the island women. But the link is hardly inevitable. The Russian academic I. S. Licevitch has remarked on the fact that, while 18th century Europeans would marvel at the technology of their 20th century descendants, they would probably be horrified by the current low level of decency and popular culture. Technological supremacy is no guarantee of cultural refinement or moral virtuousness. Conceivably, the aliens could be scientifically hi-tech and culturally low-brow.

What, then, can we say about E.T.'s ethical being? That he will have a moral code seems highly probable, given the obvious survival value that any society derives from regularizing the behavior of its members. He may also be religious. But both his moral code and his theology will serve the purposes of his own kind. His views towards us, or towards any other creatures that may exist among the stars, may be as divergent as ours are towards cockroaches and cocker spaniels.

When it comes to describing the mental and moral makeup of the aliens, the sci-fi films usually fumble the ball. Most of the Hollywood product would have us believe that E.T., once in contact with earthlings, will be interested in our societal values, our music, our kids, and possibly even our job opportunities. Wide-screen aliens may be here for good or evil, but at least we can discuss such matters with most of them. Their science and technology are sophisticated, but their intellectual acumen is not very dissimilar to our own. They either share our moral code, or are completely amoral.

Alas, for the case of any real, detected E.T., most of this cultural congruence is doubtful. He will be technologically and intellectually far in advance of us, and his culture will be far older than ours. We might be unable to understand him at any level. E.T. is likely to be ethical and possibly religious, but his attitude towards the inhabitants of other worlds is unpredictable. In the end, the only thing we might have in common is science. The laws of the universe are, not surprisingly, universal. They may be formulated in a multitude of ways, but that's not the point. If, for example, Nature allows only one efficient mode of communication between star systems—electromagnetic waves—then the unearthly way in which the aliens describe these phenomena is of little consequence for our search, although it might make for an intriguing point of discussion once contact is established. Any E.T. that we either hear or come across will have mastered science. Beyond that, we can only say that humans are undoubtedly the low-level baseline for the complexities and capabilities of the extraterrestrials. Our speculations on the details of their nature may be likened to an attempt by tanked goldfish to describe the abilities and motivations of the two-legged creatures who share the house.

MACHINE INTELLIGENCE

Our view of E.T. is heavily predicated on what *we* are: the first thinking beings to occupy the planet. We assume that if the human species can figure a way to survive its own destructive tendencies, then perhaps an extraterrestrial species has managed to do the same. The result would be the establishment of a long-lived, technological civilization. Such societies may exist elsewhere in the Galaxy, and we might either overhear or meet them. But note that we have subtly assumed that our contact will be with the *first* sentients to appear on the planet they occupy. In fact, our quest is not so restrictive. We're only looking for cosmic company. We don't insist that any aliens we discover are the original thinking critters from their own star system. All that counts is that they be around now.[6]

This is an intriguing point simply because the universe is old and the Earth is young. Other worlds will have started down the tricky paths of life billions of years ago. They may have spawned intelligent beings long before our planet came into existence. But species come, and species go. As unpleasant as the thought may be, even *thinking* species might come and go. A commonplace scenario, as we've already noted, might be one in which intelligent creatures spring up on a planet, strut their brief stuff, and are forthwith extinguished by external circumstance, or more plausibly by their own hand. Many editorial writers have postulated such a dismal end for *Homo sapiens,* after all.

---

6. In the case that we find the aliens by, say, overhearing a radio broadcast, "around now" means that they were around at the time the transmission was sent.

If this dystopian view of alien nature is accurate, then self-destruction could frequently bring quick and ugly ends to technological species (a point we will return to later). In that case, our efforts to locate intelligent neighbors might be stymied. Shortly after the aliens become sophisticated enough to build radio transmitters and rockets, they will suddenly expire in a mushroom-shaped cloud of smoke.

But if, as we have emphasized, intelligence has real survival value, it will come back like a bad penny. It will arise Phoenix-like from the ashes of its prior destruction, undoubtedly in new form. Imagine being on such a world when civilization erupts for a second time, and the archaeologists begin to dig up the ruins of a long-gone, intelligent species on their own planet! This phenomenon could be encouraging news for those seeking cosmic company. If intelligence is persistent, if sentient species routinely have successors, then the outlook for finding them brightens considerably.

Another possibility is that thinking beings can manage to avoid any such break in the reign of intelligence. They might do this by finding a cure for the built-in aggression that is an inevitable by-product of evolution in a competitive world. A civilization whose members were content to contemplate their navels or play harmless video games might last a long time. But another, less insipid approach is possible. A sentient society might be able to short-circuit biological evolution, and deliberately engineer its own successors. This seems plausible simply because we see signs of it on our own technological horizon. We may be unwittingly taking the first steps toward producing the next thinking inhabitants of this planet.

Consider the medical practice of transplanting body parts, such as hearts or knee joints. (Whether the parts come from other humans or from other species, for example by using a pig heart instead of a human one, may be of great interest to moralists, animal activists, and the pig community. It is not particularly relevant to our argument here, however.) Success with these biological transplants will create a strong drive to develop artificial replacements. Manufactured parts will, in the long run, be cheaper and more readily available when required. The recipients' bodies will also be less prone to reject them.

Dialysis machines (artificial kidneys) are a contemporary example, although today's models are too large for comfortable implantation. But there is little doubt that the next half-century will witness the development of a greater variety of both artificial body parts and synthetic bodily fluids, such as blood. At some point in the next century, we may, indeed, have the technology for the *Six Million Dollar Man*.

Parallel with this partial mechanization of humans will be the humanization of machines—the production of sophisticated robots. These devices will be outfitted with computing horsepower that will allow them to take on far more challenging work than today's models, such as those that put together the family car. Robots designed to do the tasks that humans would rather avoid—picking strawberries, mining coal, or servicing nuclear reactors—are not far in our future.

These easily-foreseen co-minglings of machine and protoplasm are, in a broader sense, simply hi-tech tool use. For thousands of years we have been constructing devices that facilitate our existence, that free us from the infirmities and unpleasant labor that interfere with what we like to do most: contemplating new physics, chasing one another around the desk, or dozing in front of the TV. We can confidently expect that within our lifetimes, the rude chipped flints of our forefathers will evolve into sophisticated household robots and implantable spare parts.

This is all but a modest extrapolation of current capability. The situation changes radically, however, if we take one more step and produce a stand-in for our brains. This would go far beyond tool use, for now we have replaced the guy using the tool. We will have engineered our successors.

Is this possible? Everyone knows that machines can't think. The best among them can play a depressingly good game of chess, but none can, say, write a book on the Zen of Chess. For years, researchers toiling away in the optimistically-named field of Artificial Intelligence have tried to build machines that could successfully tackle such relatively simple problems as stacking tin cans or diagnosing illnesses. Their success so far is modest. But the researchers involved remain upbeat in their belief in the possibility of thinking machines, although the idea is not without naysayers. Among those who are not so sanguine about the possibility of synthetic sentients is the British physicist Roger Penrose. He has argued that consciousness and creativity may depend upon unpredictable quantum mechanical processes, and would therefore be beyond the capabilities of a deliberately constructed device. However, most scientists subscribe to a more mundane and mechanistic view of our thinking apparatus. MIT physicist Philip Morrison has described brains as merely "slow-speed bit processors operating in salt water." If so, it is quite possible, in principle, to replicate their functioning in dry hardware. The detailed construction of such a machine might be somewhat different than that of the human brain, in the same way that airplanes don't imitate birds by flapping their wings. But the functional behavior could be the same.

Science fiction long ago mastered the production of such cybernetic cerebellums. A contemporary example is the android Data, a favorite character on *Star Trek: The Next Generation*. Data is human-shaped, presumably for ease of accommodation aboard the U.S.S. Enterprise-E, a ship built, and largely crewed, by hominids. He has landed the job of Operations Manager, showing that even in the 24th century, machines are still putting humans out of work. Even more perplexing, Data has been accorded full civil rights as a sentient being. This is enlightenment of a curious sort, for it suggests that if enough citizen androids were constructed, they might vote themselves into office and take over civilization via the ballot box.

Data was programmed without emotions, although an emotion upgrade chip was made, and even inserted into the android's socket. Lamentably, the chip burned up due to a power problem, and Data remains saddled with a Pinocchio complex: he wants to be human. Data longs to understand us, and even develop a sense of humor. He wants to be a good primate.

Data is like many aliens of both fiction and the popular imagination: he suffers an emotion gap. And because of this, we feel a bit sorry for him. Poor Data; he's got all the equipment any red-blooded male has, and is programmed in multiple unspecified 'techniques,' but he can't quite muster love. Naturally, this gives the android some reassuring vulnerability. But emotions, as we've noted, are useful survival tools for a biological being in a competitive environment. They could be less beneficial for a machine, and as Mr. Spock might remark, could lead to behavior that (from a machine's point of view) is "not logical." Androids don't have to get mad to defend their turf or fall in love to ensure their progeny. They might be as cool as chipped ice.

However, it is Data's intellectual prowess that is of interest here. He claims to have a mental storage capacity of 100 quadrillion bytes, which is about a hundred times as much information as you'll find in the Library of Congress. Despite this impressive total, the android doesn't shame his fellow (human) crew members with overwhelming intellectual insight and brilliant deduction. Data's just another guy on the ship's bridge (except that his jokes tend to fall flat).

But imagine a real thinking device, able to react to new situations, plan ahead, and evaluate strategies. It could be fed the world's accumulated knowledge and not forget a thing. One of the first tasks we might put before this super savant is to design its own successor—a machine more capable than itself. And that machine would be asked to do the same. In short order, we could produce a device that the slow and uncertain processes of biological evolution might never bring forth.

The consequences of silicon smarts for our own society would no doubt be revolutionary. But for the purpose of finding the extraterrestrials, the important point is that truly capable machines would be able to do something that is quite hard for biological intelligence: they could journey to the stars. Biology is fragile, and complex organisms live only a short time. Interstellar travel for biological aliens will only be practical if enormous velocities, close to the speed of light, can be attained. Machines, on the other hand, are less likely to be in a hurry. Cheaper and safer slow-speed rocketry might be an acceptable travel mode for a machine.

Why would a machine leave the planet of its birth? One obvious attraction would be the fact that its natal neighborhood might, like Earth, be in a relatively dull part of the Galaxy. When asked what the truly interesting and important things in the universe are, most humans would probably answer "sex and money." In fact, a more global answer is matter and energy (which we manage to convert to sex and money on Earth). Matter and energy can be found in far greater abundance elsewhere. For example, in the central regions of our Galaxy, the density of stars is more than a million times higher than in the fringe areas we inhabit. We are in the galactic boondocks, and a machine capable of leaving home might naturally hanker to go where the action is.

So one possibility is that machines developed billions of years ago by a distant, alien civilization have spread through the star fields of the Milky Way, perhaps producing occasional duplicates as back-ups, and are now cruising the interstellar voids like a flotilla of insects. They could rapidly adapt to the harsh environments of the Galaxy, since machines can improve themselves. This ability doesn't extend to living species, despite beliefs to the contrary by early researchers. At the beginning of the 19th century, the French biologist Jean Baptiste Lamarck proposed that creatures could influence the characteristics of their offspring by their own behavior. In straining to reach leaves on high branches, giraffes would stretch their necks, and this modification would be passed on to their progeny. In fact, acquired traits like a stretched neck are not inherited. Biological evolution, as Darwin showed, is a slow and haphazard process of pruning for comparative advantage.

But machine improvement wouldn't proceed at Darwin's languorous pace. The machines could rapidly refine themselves, and adapt to existence in the rarefied bath of starlight and gas that fills the vastness of space. Their evolution would be Lamarckian and speedy. Their ability to think might earn for them the honor of being the true intelligentsia of the Galaxy.

In characterizing the aliens, we can make plausible arguments as long as they are biological. Our description of E.T.'s cultural level, cranial capability, and moral bent are all based on traits and behaviors that have clear survival value in a competitive, life-filled environment. They obviously apply to us, and, one assumes, to any biological aliens that inhabit a planet not enormously dissimilar to Earth. But if E.T. is a machine, and possibly a lonely machine with very little daily interaction with others, then his attitude and behavior will be strange beyond our imaginings.

It may be that the ultimate achievement of biological intelligence is to get the silicon sentients underway. Biology is a nice, necessary first step, and a creature like *Homo sapiens* is an example of a simple foray by life into the realm of thinking entities. But the next big step, the truly dramatic step, is the start of machine intelligence. You might regard this scenario as both arguable and depressing. Nonetheless, if sentient machines do exist, then their obvious advantages could make them an important, if not the dominant, intelligence in the Galaxy. Should SETI scientists succeed in picking up a signal from the cosmos, no one will be surprised to learn that the signal comes from a machine, a radio transmitter. But we should also be prepared for the possibility that *that* machine is in the service of another machine.

Alien Motives

When Hollywood aliens make contact with earthlings, they inevitably have an agenda. The extraterrestrials have reasons for getting in touch, and these range from the biological to the economic. The focus of their occasionally helpful, but more often destructive intentions, is either to help themselves to our planet's real estate, or to its mineral or human resources. Prune-faced little *E.T.* is an apparent exception: his original purpose in visiting our world was to collect some specimen plants for science. But once abandoned by the other members of his field trip, *E.T.*'s interests joined the alien mainstream, and shifted from the local flora to the local folk.

E.T.'s switch from plants to people was necessitated by self-defense. The neighborhood kids were able to shield him against pernicious government bureaucrats. But in most cases where cinematic aliens tangle with humans, it's not the extraterrestrials who need the protection. The vegetable alien of *The Thing* apparently had only one thought on his leguminous brain: to drain the blood out of every human in reach. *The Blob* was similarly single-minded (if one can imagine such a gelatinous creature having a mind). He simply visited our planet as a starving student might visit a cheap buffet: for unlimited feasting. The toothy creature of *Alien* abused the human crew of a visiting spacecraft by first using them for reproduction and later for nourishment. Dinner and sex, but not necessarily in that order.

Occasionally, the celluloid extraterrestrials display a more helpful attitude. In *Cocoon,* friendly aliens rejuvenate some senior citizens, while in *The Day the Earth Stood Still* cosmic visitors land on the White House lawn to spout off about how we can all get along in the Galaxy. Steven Spielberg, who evidently prefers benign extraterrestrials, brings a mother ship-full of friendly cosmic critters to Devil's Tower, Wyoming in *Close Encounters of the Third Kind.* They ultimately whisk away the film's human protagonist, presumably to give him a cheerful tour of their home star system.

Despite such occasional meddling in human affairs, most Hollywood aliens come here for the real estate, and humans are merely an annoyance to be evicted or exterminated. The first and foremost example is *War of the Worlds,* in which the Martians elect to move from their planet to ours. The same general idea inspired the invading extraterrestrials in *The Arrival,* who busily set about re-engineering our atmosphere to better suit their gusto-grabbing, alien lifestyle. *The Man Who Fell to Earth* drops in because he wants our planet's water, and the repugnant extraterrestrials of *Independence Day* pay a visit only to strip the world of its resources and move on.

Earth is evidently the tourist mecca of the Galaxy. But do we really have that much to offer? Will the aliens be tempted to get in touch, either because they want to plunder our planet or to use our bodies for snacking or breeding? Do they have something urgent to tell us that they feel we must hear?

These questions enjoy an importance beyond the demands of Hollywood science fiction. Our attempts to establish the presence of intelligent extraterrestrials, let alone extraterrestrial contact, requires initiative on *their* part. That's obviously true if we expect them to visit our solar system, although few researchers in the field expect the aliens to show up in person. But even the detection of a radio signal might only occur if distant societies deliberately beam broadcasts in our direction.

Consequently, it is a fair and relevant question to ask what might motivate the extraterrestrials to reach out and touch someone. If there is no reason for them to do so, their existence may never be discerned.

THEY WANT OUR BODIES

As we've noted, many motives that would drive aliens to get in touch, often in the literal sense, have been explored by Hollywood. Film-makers usually feel obliged to give the extraterrestrials some halfway-plausible reason for making what must be a taxing and tedious trip from the stars. Not surprisingly, most of the Hollywood motives are strongly anthropocentric. After all, audiences can readily identify with alien impulses that are, in fact, merely transposed human impulses. And what could be more anthropocentric than bodily functions?

Among these, eating is at the top of the list. Humans, the product of nearly four billion years of evolution on an eat-or-be-eaten world, have had experience with this. There aren't any other earthly species that seriously try to breed with *Homo sapiens,* but more than a few will gladly turn us into a meal. Might aliens do the same?

One obvious deterrent to feeding on far-off neighbors is the required travel. It's highly implausible that extraterrestrials would traverse the daunting distances of interstellar space just for dinner. The energy required for the trip would dwarf the cost of producing food on their home planet. For a 100-pound alien to come to Earth (no rocket, just the alien) at even the modest speed of one-tenth that of light would take as much energy as burning a billion gallons of gasoline. A hungry alien could certainly produce plenty of gourmet meals at home for far less than this impressive expenditure.

But suppose the extraterrestrials were here only accidentally. Perhaps the voracious mound of red Jell-O that confronted Steve McQueen in *The Blob* didn't actually intend to make a gelatinous touchdown on our planet. Maybe he only inadvertently plopped to Earth, and was compelled to forage in order to survive. The same inconvenience may have befallen *The Thing,* the vegetable alien thrown from a saucer while making a crash landing in the arctic ice, or even the *Alien,* who, after all, had to eat *something* in the spacecraft where he was born. Does it make sense that in such circumstances visiting extraterrestrials would elect to chow down on the protoplasm at the top of the food chain, namely us?

Not likely. The purpose of eating, other than the social benefits of pleasing your mother or entertaining a date, is to provide your body with the chemical compounds necessary to supply it with both energy and the materials for building new cells. For earthly organisms, food-stuffs consist of carbohydrates, fats, and proteins. Nothing else ingested on this planet is digested. But would E.T. have the guts for such delicacies?

We've noted earlier that E.T.'s biochemistry will likely be carbon-based, and consequently the molecular building blocks of his body might be similar to our own. Consider glucose, for example, a simple sugar and one of the products of photosynthesis. It is the basic energy source for earthly life and, one could argue, is such a straightforward molecular construction that it might conceivably be a universal "power bar." The aliens might have a taste for glucose.

If so, they'll be tempted to grab for the fruit bowl, and will probably leave us alone. Humans aren't interesting reservoirs of simple sugars. They convert these basic energy compounds into complex molecules called fats, and then store them away in thighs and paunches. An alien that took a bite from a beer belly would get a mouth full of gristle, but literally wouldn't have the stomach to digest it. His gut would lack the specific enzymes required to convert this fatty snack into simple sugars.

Perhaps the aliens are not keen on saturated fats anyway. Would they be lured by the thought of protein-rich human steaks? Possibly, but an impediment lies in the curious fact that the proteins of all life on Earth are composed of chains of amino acids that are invariably left-handed. When amino acids are formulated in a chemistry lab, they are produced in equal numbers of right-handed and left-handed versions. This 'handedness,' or chirality, refers to the asymmetry of their molecular structure. For reasons that are still unclear, all earthly life uses left-handed amino acids. If E.T.'s proteins are built of right-handed amino acids, he will get no more benefit from a plate full of meat than from a plate full of sand. And even aside from the problems of chirality, it's overwhelmingly likely that biochemical incompatibility would preclude a random, alien creature from making use of the very specific animal proteins found in human steaks.

The ability of alien digestive systems to metabolize sophisticated earthly foodstuffs is doubtful, and that limitation may even extend to the simple carbohydrates, such as the sugars mentioned earlier. The most abundant complex carbohydrate on Earth is cellulose, the principal stiffener of plants and houses (in the form of two-by-fours). Cellulose is everywhere, and has been around a long time. Nonetheless, humans, who evolved on this planet, are unable to make a nourishing meal out of lumber, despite its abundance. Even we can't digest this plentiful carbohydrate.

Our biochemical requirements and digestive equipment have been finely tuned by almost four billion years of sharing the planet with the plants and animals upon which we dine. Our needs and our gastric chemistries are highly specific. E.T. won't have the guts for earthly edibles. For him, they will be empty calories.

Well, what about that other motivation perennially popular in Hollywood—reproduction? Moviegoers groaned audibly when John Hurt gave birth via spontaneous Cesarean section to an unattractive extraterrestrial in *Alien*. Procreation was also the name of the aliens' game in the *Invasion of the Body Snatchers*. Sinister pods left in earthly neighborhoods like abandoned cars slowly replicated the physical form of the neighbors while eliminating the originals. The pods' imitation hominids looked great, although they had the emotionless, bland personality that so many Hollywood aliens are condemned to endure. The same was true for humans assimilated by the vacuous Borg, the half-biological, half-android creatures of *Star Trek: First Contact*. In these cases and many others, the aliens find us a desirable and expendable template for reproduction.

In an earlier chapter we noted the fact that we won't share DNA (let alone, *human* DNA) with extraterrestrials. It's also inconceivable that all the other specific paraphernalia of mammalian reproduction, from placentas to hormones, would be compatible with the spawning of aliens. But what about the possibility that we are only being used as a body plan, as occurs in several of the films cited? Doesn't it benefit the Borg to meld the best characteristics of humans with hi-tech hardware?

Probably not. As Mr. Spock would no doubt intone, "It isn't logical."

Imagine a situation a few hundred years hence, when our knowledge of molecular biology has reached such a level that we are able to graft DNA from one species to another, and thereby combine desirable characteristics from different animals. Would we be tempted to incorporate the eyes of a hawk in our kids? The echo-location ears of a bat? Aside from the peculiar effects this might have on our appearance, it's a totally wrong-headed approach. If our technical skills permitted such things, they would also allow us to simply re-engineer human eyes and ears. And we could always do the sort of thing we do now: build machines; make miniaturized night-vision scopes and portable sonar equipment. We wouldn't consider crossing humans with cheetahs to speed transport. We would construct hot cars instead.

The same sort of argument can be made against a slightly less direct reproductive strategy, the idea of using humans as a surrogate "nest" for growing aliens. While this arrangement might provide a little body heat, the biological incompatibilities are formidable, and the available nutrients might not be exactly what an extraterrestrial fetus needs. If the extraterrestrials can surmount these problems to incubation in humans, then they can surely devise their own sort of *in vitro* incubation at home. Any extraterrestrial species that involves humans in their reproduction has long ago gone extinct.

This fact is not unrecognized. In 1997, a British insurance broker, Goodfellow Rebecca Ingrams Pearson Ltd., boldly declared themselves the first to offer insurance against impregnation by aliens. For 100 pounds a year, skittish British could protect themselves (at least financially) against extraterrestrial pregnancy. If the worst occurred, the insured would be compensated 200,000 pounds.

Even if we share biochemistry with the aliens, we would certainly not share the *specifics* of our biology. When Charles Darwin landed in the Galapagos Islands a century ago, he found the local iguanas to be of some interest. However, he didn't try to breed with them nor (so far as is known) turn them into a meal. Real aliens will not want humans for either purpose.

NATURAL RESOURCES

Perhaps E.T. has a hankering for our resources. David Bowie, as *The Man Who Fell to Earth,* portrayed an extraterrestrial with a thirst for our planet's abundant water. He wasn't the first fictional alien to covet liquid assets, of course. H. G. Wells' Martian invaders in *The War of the Worlds,* the prototypical hostile visitors from space, were motivated to leave home because of their planet's lack of moisture.

Wells' story was a straightforward elaboration of the claims by astronomer Percival Lowell. Lowell, as we've noted, believed that Schiaparelli's *canali* were, in fact, designed to irrigate a dying, drying planet. The Martians were in desperate circumstances, undergoing a planet-wide water crisis. Shortly after Lowell publicized these ideas in his book *Mars,* novelist Wells decided that the desiccated Martians might recognize a far more expedient solution to their problem: "Looking across space with instruments, and intelligences such as we have scarcely dreamed of, they see, at its nearest distance only 35,000,000 of miles sunward of them, a morning star of hope, our own warmer planet, green with vegetation and gray with water, with a cloudy atmosphere eloquent of fertility, with glimpses through its drifting cloud wisps of broad stretches of populous country and narrow, navy-crowded seas." According to Wells, the Martians would soon tire of the expensive civil engineering works required to solve their problems at home. Better to pull up stakes, pile into the rockets, and move to a wetter neighborhood.

Could it be that real extraterrestrials would find our watery world attractive?

No. In the case of the putative Martians, they could easily have spared themselves the rigors of a trip to Earth, not to mention the expense of leveling our cities. There's plenty of water on Mars, it's just locked in the permafrost as ice. Heating up the Martian atmosphere by pumping chlorofluorocarbons into it (something we try to avoid doing on Earth) would soon supply plenty of liquid refreshment for thirsty inhabitants of the red planet. According to the best estimates, if you were to melt all of the Martian permafrost, the red planet would be covered with an ocean hundreds of feet deep, on average.

The suggestion that aliens from afar might come to Earth to *remove* our water is equally hard to swallow. Water is heavy, and therefore expensive to transport. This point was underscored when, in 1996, evidence was discovered for a field of ice near the Moon's south pole. Such a resource could someday be a boon to those colonizing our nearby satellite, simply because it may eliminate the necessity of sending large amounts of this life-sustaining liquid from Earth. And if winging water to the Moon is expensive, imagine the cost of sending it to even the nearest star, which is 100 million times farther.

In addition, if there's one compound that's undoubtedly in good supply in any decent solar system, it's water. Consider the abundance of this wet stuff in our own neighborhood. Earth has aqua a-plenty. So does Mars, although it's underground and frozen. Jupiter's moon Europa is thought to be coddled in a 60-mile deep ocean, capped with ice. That's an amount of water comparable to what is found on Earth, and astute moviegoers might ask why David Bowie didn't go to Europa to slake his home planet's thirst—especially as that moon's gravitational tug is only about 1/3rd that of Earth's, thereby reducing his transport costs. If that's not enough liquid refreshment, there are always the outer, giant planets, Jupiter, Saturn, Uranus and Neptune. All are believed to have copious amounts of water in their atmospheres.

In fact, if you look at the universe as a whole, you find that the cosmic abundance of elements favors the idea of water, water everywhere. The relative number of atoms for the top dozen elements throughout the cosmos is:

H	86.7%	Mg	0.003%
He	13.1%	Si	0.003%
O	0.08%	S	0.001%
Ne	0.05%	Ar	0.0005%
C	0.03%	Fe	0.0005%
N	0.008%	P	0.00003%

So hydrogen and oxygen are respectively the first and third most prevalent elements in the universe. Hydrogen was produced during the first few moments of the Big Bang. Oxygen has been cooked up in stars the size of the Sun and larger for a dozen billion years. With these two elements in such great abundance, water will indeed be plentiful.

But of course, Earth offers more than water. The alien critters who came calling in *Independence Day* were described as interstellar locusts who would "strip the planet and move on." What might they want to strip?

A little astrophysics can provide some insight. If you give the table of cosmic abundances a second glance, you'll notice that the most common elements in the cosmos are also the lightest. In particular, atoms that are heavier than iron are in short supply. The reason for this is straightforward. Stars run on nuclear fusion; they generate energy by melding light elements into heavier ones. For example, all stars begin their luminous lives by slamming hydrogen nuclei together to make helium. But the helium nucleus is 0.4% less massive than the hydrogen nuclei that were used to create it. The ingredients are slightly more massive than the final product. Despite what you might have learned in high school, matter can be created or destroyed. The small amount of material destroyed by fusion in the guts of stars is converted into energy.

Cooking light elements into heavier ones is the life work of stars. Their efforts always liberate energy, as long as the ingredients are relatively light. But if they attempt to fuse atoms as heavy as iron, the result is *more* massive than the ingredients. This *requires* energy, rather than generating it. No star will do this, except under some very extraordinary circumstances. Those exceptional conditions occur in the brief period when a large star runs out of nuclear fuel and dies. The star undergoes a titanic death rattle known as a supernova, and blows itself apart with staggering ferocity. In the searing furnace of this massive explosion, enough energy is available to fuse iron into the more ponderous elements.

It should be no surprise, then, that heavy elements such as lead, tungsten, and gold are in far shorter supply than light ones such as carbon, silicon, and oxygen. Stars spend their entire lives cooking up light elements. They produce the heavy ones only during the dramatic few moments of a supernova. Heavy elements are rare in the universe.

Rocky planets, such as Earth, are enhanced sources of these rare, raw materials, simply because a nearby Sun has blown away the lighter elements during their birth. It's a cosmic fact that our planet is a ball-shaped vein of high grade ore. Perhaps it would appeal to extraterrestrial prospectors?

In fact, it probably wouldn't. To begin with, there are diggings in the solar system that are more attractive. The asteroids are rich in heavy metals, and because of their size and low gravity, are far easier to steal and feed to a smelter. For years, Michael Pappagiannis, an astronomer at Boston University, suggested that extraterrestrials might be interested in quarrying our asteroid belt. But prospecting by aliens anywhere in the solar system is a doubtful proposition. The major hurdle for interstellar miners is, once again, the transport cost. The energy required to accelerate a ton of gold or tungsten to, say, one-tenth the speed of light (to enable delivery to nearby stars within a century or two) is more than the energy required to fuse iron and an assortment of light elements into these heavy metals. The aliens could *make* gold for less energy cost than bringing it back from Earth.

THEY WANT TO IMPROVE OUR MINDS OR OUR MORALS

Food, sex, financial gain. We can readily envision these motives for alien contact because they are familiar incentives for many earthly interactions. But could it be that the aliens have moved beyond the carnal and the commercial? Might they want to get in touch for more altruistic reasons?

In their book *Intelligent Life in the Universe,* Iosef Shklovskii and Carl Sagan remark on the fact that much of the early colonization on Earth was motivated by a desire to spread the Christian faith. Perhaps the aliens would come here to proselytize, or alternatively opt to stay where they are, but indulge in high-powered, broadcast evangelism.

We've already noted the ubiquitous presence of religion in earthly societies, so it would not be unexpected for an alien civilization to be religious also. Spreading the faith is certainly one reason they might take the technological initiative required to reach out.

Hollywood aliens have not shown much penchant for religious evangelism, but they occasionally stop by our planet to give us a moral lesson or spout philosophy when we visit theirs. In the 1951 film, *The Day the Earth Stood Still,* the alien Klaatu, played by Michael Rennie, is on a mission to warn humanity. Evidently alarmed by the machinations of the Cold War, the extraterrestrials, like Mom, have decided we need a good talking to. So Rennie lands his saucer near the White House, and lets loose with some stern pronouncements.

"It is no concern of ours how you run your own planet," intones the gentlemanly, if disingenuous Klaatu. "But if you threaten to extend your violence, this earth of yours will be reduced to a burned-out cinder. Your choice is simple: join us and live in peace or pursue your present course and face obliteration. We shall be waiting for your answer. The decision rests with you."

Having delivered his message to behave or else, Klaatu rockets back into space. But he leaves behind Gort, an eight-foot-high-at-the-shoulder robot, as enforcer. Gort's job is to make sure we take Klaatu's advice. While a warning from the aliens about the dangers of nuclear war might sound beneficial, the idea that they would use their technology to control our behavior relegates us to the level of pets. In the treatment of American natives by missionaries of the 18th century, we see a similar, and disheartening mixture of noble intent and curtailment of freedom. The Indians were inducted into the Christian faith, and then told to work the mission's farms. Aliens bearing moral messages might not be such a good thing.

But irrespective of whether we would benefit from such efforts to improve our behavior, it seems unlikely the extraterrestrials would make them. Human morality is something we apply only to our own species, after all. Man must behave; animals are free to do their own thing. The thought that the aliens—who are most assuredly an entirely different species—would take an interest in changing our personal attitudes is akin to suggesting that we might try to improve the moral character of goats. Unlike the premise in *The Day the Earth Stood Still,* we are no threat to any extraterrestrial civilization that is able to commute between the stars. Such advanced aliens might have some academic interest in studying our behavior, but rather little incentive to change it.

WILL THEY BE AGGRESSIVE?

Aside from possible moral agendas, to what extent would the aliens regard us benignly? As mentioned, movie director Steven Spielberg's aliens are always friendly. *E.T.* is positively childlike, and the diminutive, white aliens of *Close Encounters of the Third Kind* were pleased to offer a saucer-load of humans a ride to their world. In *Cocoon,* visiting extraterrestrials take time out from a hectic schedule to rejuvenate some senior citizens.

It's hard to imagine a society of highly advanced creatures taking such a personal interest in us, although in 1975, a Florida politician named Lynne Plaskett claimed that aliens cured her cancer. When we visit Antarctica, we don't offer the penguins rides on our boats, cure their infirmities, or spend a lot of time either playing with their juveniles or their aged. Even aside from the question of specific interest in human activities, would the aliens be as gentle with a foreign world and its inhabitants as these scenarios suggest? Would they be non-aggressive because that is a more perfect state of being?

Astronomer Sebastian von Hoerner is one of many researchers who argue that the aliens will, indeed, be beyond aggression. The growth in destructive capability of earthly technology convinces von Hoerner that any civilization that has survived its own scientific evolution will have managed to engineer aggression out of its system. The only long-lived extraterrestrial societies will be those that are peaceful and passive.

This is a nice thought, and seems to vindicate Spielberg's kindly view of aliens. However, if all advanced extraterrestrials are content to play video games or contemplate their navels, then we are very unlikely to either hear them on the radio or find them in the neighborhood.

Interstellar space travel is risky. The broadcast of strong signals is expensive and possibly dangerous. Passive aliens may not undertake either. So if all advanced societies manage to rein in their aggressive tendencies, we might never know of their existence.

However, aggression has its good points. Despite some occasional regrettable consequences, taking the initiative has survival value. In addition, a society composed of uniformly placid beings could be wrecked by an errant cosmic ray that accidentally resulted in the birth of an individual with more than the usual get up and go, with the "killer instinct" that will garner him a few hundred girlfriends and plenty of power. Within a generation, aggression would be back in the system. It may be tough to keep aggression down.

When it comes to getting in touch, the historical record of our own planet tells a very one-sided tale about which creatures make contact. The Incas of Peru didn't meet your average, affable man-in-the-street Spaniard. They were treated to the go-getting, ambitious conquistador, eager to further his career and make money for his sponsors. The passive guys stayed home. The fellows at the front door were the aggressive ones.

It is inevitable that any aliens who take the trouble to either signal their presence or transport themselves beyond the bounds of their own solar system will be, by definition, aggressive. The assumption that there are many such societies is underscored by the survival value of aggressive behavior. On the other hand, we've considered some of the most commonly offered inducements for aliens, aggressive or other-wise, to visit our planet, and found them all weak. They may be aggressive, but not inspired to travel. So will extraterrestrial contact ever involve interstellar rocketry? Is there any motive for the aliens to make the arduous trek from one star system to another?

There might be one. They may simply want to spread out.

COLONIZERS

In the 1996 film *The Arrival,* human society is bulldozed out of the way while invading aliens reshape Earth to accommodate their own environmental requirements. The extraterrestrials have landed, and are moving in.

As is the case for most Hollywood motives, the idea that the aliens want our turf is a stretch. It's true that good planets may be, if not few in number, at least far between. But how much mileage are you willing to rack up to take over another planet? Consider what we might do. The single truly decent world in our solar system is Earth, of course. And there's little doubt that Earth is filling up. Depending on your tolerance for neighbors, you might be of the opinion that our planet is already near to its human carrying capacity. So we, too, may soon wish to increase our *Lebensraum*. Would we rocket off to another star system to find a close analog of Earth, exterminate any interfering locals, and spiff up the atmosphere or anything else as necessary to suit our needs?

It's not inconceivable, but at first blush this idea is in the same league as building a nuclear reactor to power a toaster. If we want more real estate, it can be found far closer than the stars. The moon is an obvious enticement, less than a day's journey distant. Mars offers water and an atmosphere, and is likewise nearby. Colonization of these adjacent worlds would double the amount of acreage available to earthlings. But if we *really* want to expand our habitat, we should consider the ideas of Gerard O'Neill and Thomas Heppenheimer. In the 1970s, these two physicists described how we might construct giant aluminum cylinders that would be parked in orbit near the Earth-moon system. Artificial space habitats, rotating slowly to mimic gravity, could house millions of people in comfort and style. The bulk of the raw materials needed for construction would be catapulted off the low-gravity moon, rather than being rocketed at great cost from Earth. It was anticipated that, by the 1990s, millions of humans would be living in such aluminum condo cans.

That hasn't happened, but the reasons are economic, rather than technical. In the next century, we will probably build artificial space colonies. Their capacity to accommodate a proliferating society is enormous. The same is true of schemes that would convert the myriad asteroids of our solar system to inhabited oases. Princeton physicist

Freeman Dyson has pointed out that the amount of real estate available on the asteroids is ten thousand times what we have on Earth. The asteroids also sport the elements necessary for breathing, drinking, and construction.

In truth, it is much simpler to expand throughout your own solar system than to go thousands of times farther away and colonize someone else's. But even so, it might still be done. It would have been logistically simpler for the early American settlers to remain on the continent of their birth, and found new societies in Europe. But persecution or wanderlust drove the pilgrims to undertake dangerous and long journeys. Their destinations were inhospitable and their fate unsure. All that beckoned them was hope for a better life. This gamble didn't appeal to many; most Europeans stayed where they were. But those that shipped out eventually colonized a hemisphere. If the uninvited aliens of *The Arrival* are emigrants, they might have reasons to travel that are sufficiently compelling to take on the difficulty of traversing the light-years. The fact that someone's already at home on Earth would likely deter them no more than the presence of American Indians forestalled the Europeans. Accommodate the natives, if possible. Learn from them, surely. But when it comes to appropriating real estate, prior habitation is no real impediment. The aliens of *The Arrival* got on with their bulldozing.

Colonization is difficult to rule out as a motive for interstellar dispersal because it is done for many reasons. These range from the ephemeral (a desire to homestead new territory) to the eminently practical (the impending death of your sun). The possibility that some galactic civilizations have already made a colonization effort leads to a peculiar problem, one that has divided researchers in the field of extraterrestrial intelligence into camps.

Imagine that at some point during the history of the Galaxy, an alien society rockets a contingent of its populace to a nearby star. There are many reasons why they might leave, but it is the *process* we are considering now. The speed with which they venture into space is unknown of course, but for the sake of discussion, let's assume it to be 1% the speed of light (our own rockets are far slower, only 0.005% light speed, but these space-faring aliens are better technologists than we). The alien colonists will reach the nearest star systems in roughly 500 years. Suppose that their efforts to set up a homestead are successful, and that after another 500 years, the new settlements are so populous (and to some, so unappealing), they spawn their *own* émigrés. The new colonists spend 500 years reaching the nearest stars to *them,* and so on. The "wave of colonization" expands at 0.5% the speed of light, since we've assumed that it takes as long to build up a new colony as it took to reach it.

Since the Galaxy is roughly 100,000 light-years across, this colonization effort—assuming it is sustained—will reach every part of the Milky Way in 20 million years or less. While 20 million years sounds impressive, it's an inconsequential length of time compared to the Galaxy's age, which is closer to 10,000 million years. This argument is not greatly affected if we change the speed or colonization time. And what our simple calculation shows is that, if any civilization was ever keen to colonize the Galaxy, they've had more than enough time to do it. It is somewhat akin to what happened after Columbus' discovery of America in 1492. Within thirty years there were Spaniards up and down the coasts of the New World. Once underway, the process of colonization is fast.

Consequently, if advanced civilizations have sprung up at some early time in the Milky Way's long history, they should be everywhere by now. We would expect to find ubiquitous evidence of their activities. The same would be true if intelligent, self-replicating robots were doing the spreading out. If any society has ever launched either themselves or clever robots into space, the Galaxy (including our own small niche) should now be chock-a-block with alien intelligence. As far as we can tell, it isn't. Since neither colonizing critters nor intelligent robots seem to be hanging out in the neighborhood, does this mean that we are the only sentient beings around? In the summer of 1950, the eminent physicist, Enrico Fermi, recognized the paradox of this conflict between expectation and local observation during a lunch at the Los Alamos National Labs, and asked, "Where is everybody?"

Where indeed? Was this a serious problem or not? In 1975,

astronomer Michael Hart published a short discussion of Fermi's paradox and concluded that galactic colonization is so likely to have occurred that the lack of aliens on our doorstep is significant. Hart raised the dander of many by emphasizing that the absence of extraterrestrials in one place—namely, Earth—could mean that we share the Galaxy, and possibly the universe, with *no* one.

The provocative gauntlet thrown down by Hart was soon joined by one from Louisiana physicist Frank Tipler. Tipler felt that Hart could go farther. Even if the aliens stay home themselves, they'd surely send out a few replicating robots, argued Tipler. The machines would multiply and rapidly infest the Galaxy. They would alight on every decent world, including ours, and leave no planetary habitat unspoiled. But Earth *is* unspoiled, at least by extraterrestrials, and the same seems to be true of the moon and Mars. Tipler's conclusion: humans are the Galaxy's best and brightest.

This is, of course, a serious affront to the Principle of Mediocrity. Consequently, Fermi's paradox as interpreted by Hart, Tipler and a few others has generated an energetic response by scientists who take the opposite view—namely, that despite the fact that we see neither hide nor tentacle of the extraterrestrials, they most assuredly exist. We are not the smartest things in a galaxy of a half-trillion stars. These researchers have founded a small cottage industry, churning out arguments for why the aliens are out there, but not here.

Their explanations range from the technological to the sociological. Perhaps scientifically literate societies inevitably self-destruct before manifesting their will to wander. The developing aliens invent rockets, but then use them to pulverize their own populace. They don't survive long enough to colonize the stars. Or maybe they *do* start colonizing, but the expansion effort quickly runs out of steam. Anthropologist Ben Finney has pointed out some spectacular examples of stalled imperialism on Earth. In the 15th century, the Chinese built a fleet of sailing ships for exploration and possible conquest. Only a few decades years later, a new regime took over the Chinese capital, decided that an inward-directed society was a better society, and had the ships burned on the beaches. Expansionism was caught up short by a change in management.

The Polynesians successfully spread across the South Pacific during the course of a millennium. Their island-hopping expansion is an apt analog to space colonization. But the Polynesians lost their oomph before they could colonize the coasts of the Americas. Their colonized islands, once idyllic, turned ugly under the twin pressures of population expansion and environmental degradation. Maybe alien expansionists have all lost the interest or ability to take over more real estate before reaching the shores of our solar system. Yet another possibility may be that the centers of extraterrestrial empire sicken and die, enticing colonists at the fringes to return to the ancient worlds their predecessors left behind. This could slow the expansion of the colonization frontier, and once again, might account for the aliens' apparent failure to come hither from yon.

As appealing as such scenarios may be, they all suffer a common failing. It isn't that we can resolve the Fermi paradox by arguing that *most* alien societies self-destruct or lose interest in expansion. *Every single one* of them must do so, for otherwise representatives of at least one society would be in our neighborhood.

This loophole, this insistence on 100% compliance, buoys the small camp of scientists who prefer to believe that the Galaxy is devoid of cosmic company. The other side has responded by imagining explanations for the absence of aliens that don't depend quite so heavily on sociological phenomena. Perhaps various advanced civilizations have divided the Galaxy into spheres of influence, and we are in "no alien's land." Or possibly our part of the Milky Way is uninteresting and uninhabited (by aliens) in the same way that central Nevada is. No one would dispute that the United States is fully occupied, and yet there are plenty of places where not a soul can be seen. America is urbanized. Maybe the Galaxy is as well, and we're in one of the uninteresting deserts.

Harvard's John Ball has taken the opposite tack, suggesting that maybe Earth is *very* interesting. Our planet might be a zoo, a designated nature preserve to be left alone. The extraterrestrials stay inconspicuously on their side of the bars.

While a cagey suggestion, the zoo hypothesis is somewhat contrived.

Most scientists who favor the idea of a populous galaxy point to a more straightforward reason why the aliens are not littering our landscape: interstellar travel is simply too costly. We've already remarked on this in regard to extraterrestrials who might be lured here by the thought of dining on humans. The meal's not worth the trip. But the argument may apply equally well to alien expansion for *any* reason. Astronomer Frank Drake points out that the energy cost for sending colonists to the stars would more than pay for an impressively regal lifestyle on their home planet. He and others say the extraterrestrials simply won't do it. Tipler rejoins that the energy costs for machines to make the trip can be far lower, because the machines are in no hurry.

The battle goes on, although there's little prospect of a winner. Of course, there is one resolution to this paradox that, on the face of it, would satisfy both camps in the Fermi feud: the aliens *are* here. Could this be so?

Getting in Touch

Fermi's question, "Where is everybody?", has prompted considerable head scratching and elicited some clever replies. But there are two answers to Fermi's query that are easy to give and disturbing to hear. One response is that offered by Hart and Tipler. The aliens are not here because they're not *there*. We are alone, and Earthlings are the most advanced creatures in an island universe of a half-trillion stars. Deal with it.

The other *easy* answer is that the extraterrestrials *are* here. They are busily buzzing the countryside in their saucer-shaped craft while occasionally taking time out to carve crop circles in British wheat or to abduct humans for some cross-species sex play.

The idea that Unidentified Flying Objects (UFOs) are from distant worlds is taken seriously by the American public. In a *Newsweek* poll made in 1996, nearly 50 percent of the respondents said they believe that UFOs are extraterrestrial spaceships. The same percentage also feels that the government is keeping this important news under wraps. This astounding result has been repeated in one survey after another since the mid-1960s. Polls have been made of the general public, teenagers, Ph.D.s, high-school graduates, and the self-styled intellectual elite of Mensa. For every category, half or more of those queried are convinced that the UFOs are not of this Earth.

Could 100 million Americans be wrong?

Well yes, they could. After all, a sizable fraction of American adults also believes in ghosts and angels. The scientific evidence for either category is minimal. But unlike either ghosts or angels, UFOs have been the subject of official government investigation. Surely there is something afoot?

SAUCER MANIA

The flood of modern UFO sightings began a half-century ago. In June, 1947, Kenneth Arnold, a civilian pilot, was flying near the Cascade Mountains in Washington state. Peering through the cockpit window, he was startled to see nine brightly glowing craft, streaking across the sky at nearly ten times his own speed. Arnold said that these craft moved erratically, "like a saucer if you skip it across water." His amazing claim was reported by Bill Bequette, of the United Press, who confounded the motion and appearance of the objects by describing them as "flying saucers." Despite this journalistic error, or more likely because of it, the majority of the thousands of UFOs seen since 1947 have reputedly had a hubcap shape. (It bears noting that saucer-shaped aircraft, once tried experimentally by the Canadians, proved not to be very airworthy in Earth's thick atmosphere. Boeing does not make flat, round aircraft. The saucers' streamlined design is also unnecessary and inefficient for travel through the airless depths of space.)

These early reports of UFOs took place shortly after the end of World War II, when large rockets offering the promise of space travel were making their debut. Wernher von Braun talked excitedly about reaching for the stars. Since exploration of other worlds finally seemed attainable for us, perhaps the aliens were already hard at it? This was the romantic view of the UFOs, and it was encouraged by a popular 1949 article in *True* magazine by former Marine Corps officer Donald Keyhoe entitled "The Flying Saucers are Real."

But while the American public was being enticed by the possibility that the ships could be visitors from other planets, the American military was working another angle. The defense establishment's concern was that these putative craft might be advanced Soviet aircraft or missiles. The Air Force put together a small group at Wright-Patterson Air Force Base, in Dayton, Ohio to keep tabs on the UFOs. This effort became known as Project Blue Book.

Even the Central Intelligence Agency got into the act. In the early 1950s, the Agency quietly assembled a small panel of highly regarded physicists and astronomers under the leadership of H. P. Robertson, of the California Institute of Technology. The panel spent weeks reviewing the 75 most convincing UFO cases collected by the Air Force. After poring through this evidence, the Robertson Panel recognized that (1) those who report UFOs do witness *something*. They are seldom pranksters. (2) Nearly all UFOs are seen against the sky (rather than, say, against a nearby mountain range), so that the crafts' size, distance, and speed are very hard to judge, and (3) the UFOs never leave any physical evidence—something you could pick up and take to a lab for analysis.

The Panel also noted that for nearly all of the sightings, a plausible explanation could be made in terms of well-known natural phenomena, friendly aircraft or balloons. Finally, in 1953 the panel concluded "that there is no residuum of cases which indicates phenomena which are attributable to foreign artifacts capable of hostile acts, and that there is no evidence that the phenomena indicate a need for the revision of current scientific concepts."

UFOs didn't seem to present a security threat to the United States, but the Air Force continued to collect the reports. By the mid-1960s, large segments of the public had come to believe that the government knew far more about alien spacecraft than it was telling. The Vietnam War was heating up and distrust of the government was rising. As a response to the growing cynicism, the Air Force asked the University of Colorado to make an independent UFO study. This effort was headed by the respected physicist Edward U. Condon.

Three years later, in 1969, the university published the so-called Condon Report. Its findings were no different than those of the Robertson Panel, convened more than a dozen years earlier. Condon said that "nothing has come from the study of UFOs in the past 21 years that has added to scientific knowledge..." Another panel convened by the National Academy of Sciences agreed. The Air Force shut down its Project Blue Book—UFOs were no longer to be investigated, even at the very low level that had been the case since the late 1940s.

Needless to say, the sightings continued, as did the charges of a cover-up. In the last decade, saucer sightings have been augmented by tales of abductions and those peculiar manifestations of alien activity known as crop circles.

Imagine the following: You are on the fourth planet in orbit about the star Zork XI, engaged in a parliamentary debate. A fellow parliamentarian has just proposed the expenditure of 100 billion galactic guineas to build a small fleet of interstellar spacecraft whose purpose is to fly dozens of light-years to an alien planet called Earth and then carve some simple designs in their wheat fields. Would you vote for this?

Probably not. And it's unlikely the extraterrestrials would, either. Yet large numbers of intelligent Britons believed that the snazzy patterns defacing their agriculture which first appeared in the late 1970s were the handiwork of artistic aliens. Some have continued to feel this way despite the fact that two retired gentlemen, Doug Bowers and Dave Chorley, emerged from a pub one day stating that they had made the "circles" with boards and rope. They demonstrated how they could create the patterns given this simple equipment, the cover of darkness, and a few hours' time. Some of these obviously hoaxed crop circles were shown to "experts," who somberly intoned that they were clearly the handiwork of extraterrestrials. As if this weren't enough, British television crews set up low-light video cameras on hills overlooking virgin wheat fields, and caught several other pranksters at work carving up new "alien designs" in the dead of night.

Many of those originally convinced of the extraterrestrial origin of crop circles have changed their mind. But not all. Those who still believe that creatures from other worlds are interested in disfiguring the back forty claim that, while many if not most of the circles are hoaxes, some are "real." This strained rationalization is similar to that used when discussing the alien nature of the UFOs. There is little disagreement, even from UFO believers, that approximately 90% of the saucer sightings have a prosaic explanation. For example, many observers mistake bright celestial bodies for spacecraft. In 1969, Jimmy Carter saw a UFO which later investigation showed to be the bright planet Venus. He was only one of many to mistake a distant planet for a nearby spaceship. Mars and Jupiter, as well as the star Capella, have fooled observers on the ground and in aircraft. Other frequent causes of UFO reports include meteors, fireballs, temperature inversions, air turbulence, Canada geese, ball lightning, aircraft, deliberate hoaxes, rocket launches, faulty radar equipment, and Japanese squidboats (the bright lights these boats use to lure their octapodal prey to the surface are so intense they even show up on nighttime satellite photos). According to Philip Klass, a long-time investigator (and skeptic) of UFO sightings, a very large number of supposedly extraterrestrial craft turn out to be advertising planes outfitted with electric signboards for flashing commercial messages.

If the substantial majority of UFOs can be readily explained as being due to such decidedly non-alien phenomena, is it not reasonable to suspect that *all* could, if there was only sufficient information?

That was the conclusion of all the government panels that looked into UFOs. Still, it must be admitted that, even if only *one* of the many UFO sightings was truly an extraterrestrial craft, the significance would be immense. The odds are long, but the payoff is high. Nonetheless, the fact remains that most scientists doubt that aliens are jockeying their spaceships through the stratosphere. One reason for this is that, despite fifty years of UFO reports, there are no artifacts. No one ever shows up with an ash tray or a tail light from a UFO. Not even a fragment. Earthlings have been sending satellites and shuttles into orbit for a mere forty years, and nearby space is littered with thousands of pieces of junk, including dead rockets, exploding bolts, a Hasselblad camera, paint chips, and the occasional astronaut glove. But somehow the aliens never experience a mishap, and never leave anything behind. (The exception to this is the celebrated Roswell, New Mexico incident of 1947, for which the government is accused of collecting every single piece of debris from a crashed saucer, as well as the bodies of its alien occupants. The fact that no clearly extraterrestrial physical evidence from this event is available for scientific scrutiny can once more be conveniently blamed on a nefarious and surprisingly efficient federal government.)

After a half-century of UFO sightings, where's the payoff? What new science do we know that we didn't find out on our own? Where's the technological spin-off, the advanced materials, the super hi-tech products that we otherwise wouldn't have? If the Air Force really has alien spacecraft under wraps at the celebrated secret test site known as Area 51 in the Nevada desert, why does our military continue to build planes following the same designs it always has? Why don't we have craft a hundred times better than anyone else's?[7] Half of all Americans think that the U.S. Government has secret information on the aliens. Is our government in cahoots with all other governments to keep this exciting information from the world? This would bespeak a degree of international cooperation not reached in any other diplomatic arena. The improbable alternative is to believe that the aliens confine their landings to America.

And finally, why aren't thousands of university researchers busy studying UFOs (in their spare time, if necessary)? There's little doubt that proof of cosmic visitors would be the discovery of the millennium. It would be a scientific coup without precedent. As one European astronomer told me, "If I thought there was a 1 percent chance that any of this UFO stuff were true, I'd spend 100 percent of my time working on it." The fact that hordes of serious scientists are not beavering away on this problem speaks volumes about its authenticity.

There's no doubt that the public likes the idea of alien visitors. *Close Encounters of the Third Kind, E. T., Independence Day* and *Men in Black* have ranked among the cinema's top box office draws. On television, *The X-Files* is a sensation. As far back as 1966, a *Look* magazine article on the purported alien abduction of just-plain-folks Betty and

---

7. Some think the aliens have provided the American military with substantial advantage. *The Day After Roswell* by Philip J. Corso and William J. Birnes (1997, Pocket Books), claims that the Army clandestinely seeded both the military and the American economy with the technology stripped from an alien craft recovered at Roswell, New Mexico in 1947. Senator Strom Thurmond wrote an enthusiastic foreword to this book, but the New York Times later reported (5 June, 1997) that Thurmond was having second thoughts. He never meant to suggest that there was any truth to the idea of alien technology being secretly disseminated by the Government, the Times stated.

Barney Hill generated unprecedented sales. It seems we have a need to believe in external, powerful personalities who can influence our lives. It's far more interesting to hear stories telling us that the aliens are here than analyses suggesting that they aren't. Philip Klass, despite painstaking research, gets precious little air time on television "documentaries" dealing with the question of UFOs. But while his message that these visions in the sky are terrestrial, celestial or hoaxed is unwelcome, it is Klass' work that best withstands the hard light of serious scrutiny.

The aliens just aren't here. Must we therefore accept the other obvious answer to Fermi's question? Must we conclude that we are without intelligent company in the Galaxy? Does the simple observation that we have Earth to ourselves compel this brutal violation of the Principle of Mediocrity, this depressingly lonely circumstance?

Of course not. We have already remarked on the host of reasons offered to reconcile a populated galaxy with a lack of aliens on our planet. Absence of evidence is certainly not evidence of absence. By the year 1491, European civilizations had been around more than long enough to have colonized every piece of earthly real estate, although they hadn't yet done so. American Indians of the time might have glanced around their local communities, seen only other Indians, and used their own version of the Fermi Paradox to conclude that they were the sole advanced society on the planet. Arguments claiming we are alone, that searches for other advanced civilizations are doomed to fail, offer no room for maneuver, no place to go. If we agree with their premise and never look for cosmic neighbors, the odds are high that we will, indeed, not find them. The premise fulfills itself. On the other hand, if we proceed to search despite the pessimism, we at least increase the chances that, if the premise is wrong, we will *prove* it wrong by finding the extraterrestrials.

Consequently, scientists involved in the search for extraterrestrials have generally adopted a "nuts and bolts" attitude, a kind of agnosticism regarding the Fermi Paradox. "Let's not worry about why the aliens aren't here. Let's build instruments capable of uncovering them in their native habitat. A pragmatic approach at least offers the chance of success." And success would be important.

PROBING SPACE

How might we find extraterrestrials? The most obvious approach is to simply go look. But as has been pointed out *ad nauseum* in this book, the stars are enormously distant. At the speed of today's rockets, a round trip voyage to even the nearest stellar systems would consume 100 thousand years and more. It is the brutal tyranny of these distances, and more precisely the prodigious energy required to bridge them in a tolerably short time, that has prompted some scientists to speculate that a voyage to the stars is only a dream. The faint lights on the nighttime sky may be forever beyond our physical grasp. In the last chapter, we remarked on astronomer Frank Drake's premise that the difficulty of interstellar travel explains the Fermi Paradox. The aliens don't build starships because of the energy costs. The clear conclusion is that we won't either.

Of course, not everyone agrees. During your next dinner party, dare to suggest that mankind will never, ever go to the stars. The response will be lively, no matter whether the guests are lay folk or rocket scientists. Since the average person has dissipated a good part of his youth watching *Star Trek* and similar television space operas, he is inclined to think that careening off to another world requires only a few dilithium crystals, warp drive, and a call to the engine room telling the people in charge to "make it so." Anyone schooled in elementary physics may retort that, dilithium crystals aside, high speed implies the expenditure of tremendous energy, not to mention the mortal danger of collisions involving even the smallest particles adrift in interstellar space. The gut reaction to such esoteric negativism is that "Yes, but in a few hundred years our technology *will* improve." After all, Christopher Columbus might have had trouble believing that, a scant five centuries after his laborious and slow trip across the Atlantic, ordinary Spaniards could jet from one continent to another in a few hours (and, incidentally, consume more energy in the process than was produced in a similar amount of time by all the oxen of 15th century Spain put together).

Indeed, our technology will improve. No one disputes that chemical rockets—our current high-tech travel mode—are simply too inefficient to reach the stars. Chemical engines derive their energy by burning stuff, a fact that's obvious to anyone watching the flame and fire of a Shuttle launch. Burning merely rearranges electrons, the lightweight particles that orbit the outskirts of atoms. But far more energy could be extracted from the nuclei of these atoms. Consider the difference in energy yield between dynamite, which rearranges electrons, and an H-bomb, which rearranges atomic nuclei. Eugene Mallove and Gregory Matloff describe in *The Starflight Handbook* practical rockets based on nuclear fission and fusion, as well as propulsion schemes that emulate *Star Trek's* matter-antimatter drives. They consider the manifold ways we might be able to construct sophisticated rockets capable of reaching, not the darkened fringes of our own solar system, but the unseen worlds around other stars. They believe that we, or our not-too-distant descendants, could cobble together one-way probes to nearby stars. The speed of these probes would be 10% that of light or less, or a few thousand times faster than existing chemical rockets. Still, they would take 50 to 100 years to make their journeys. Mallove and Matloff believe this technology is within reach, if not today, then within a half-dozen generations. The first steps are already being contemplated. NASA is considering the construction of diminutive probes that might journey a tenth of a light-year towards the stars.

Blasting off in a flaming rocket isn't the only way to go, either. Rockets, like automobiles, carry all their fuel with them. Indeed, most of the weight and space of a rocket *is* fuel. Spacecraft powered by solar sails, which would accelerate using the pressure of sunlight, could be a cheaper, if slower, means of transport. Like sailing ships of the past, they require no holds filled with combustibles. Easily foreseeable designs could reach our stellar neighbors in a handful of centuries. Another approach is to build interstellar ramjets, which would plow through space with enormous, magnetic scoops. The scoops would funnel hydrogen atoms that float between the stars into the maws of the craft's fusion engines, thereby gulping *en route* the fuel necessary for travel.

Such technologies could achieve speeds that dwarf those of the rockets we've used to explore the solar system. Building fusion rockets or solar sails in the next few centuries may or may not be practical, but it's surely not impossible. However, even if we construct these interstellar transports, they won't be nearly fast enough to permit the sort of galactic gallivanting that the cast of *Star Trek* pulls off every week.

If we really want to boldly go where no humanoid has gone before, we'll need to put the pedal to the metal. We'll need spacecraft that can reach velocities greater than half that of light. Such futuristic transports could reach Mars in ten minutes (which is not such a big deal) and nearby stars in ten years (which is). Actually, their performance is even better than this, thanks to modern physics. The laws of Special Relativity decree that when you are moving, you age somewhat less quickly than someone who's not. In ordinary life, this effect is small to the point of being entirely unimportant. For example, if you careen down the freeway for a full day at 70 miles per hour, you will age about five trillionths of a second less over that 24-hour period than your stay-at-home significant other. That's hardly worth the risk or the gas. But Special Relativity's time dilation kicks in with a vengeance at speeds approaching that of light. A crew aboard a ship traveling at 71% light speed will age only half as quickly as their relatives on the home planet. At even higher velocities, the perceived pace of time decreases rapidly. For example, people ripping through space at 99% the speed of light age at only one-seventh the rate of those who waved good-bye at the launching pad. This means that these speed demons could rocket to the nearest star, Alpha Centauri, in only six months as measured by the watches on their wrists, the number of meals eaten and games of Scrabble played, despite the fact that 4.5 years will have elapsed on Earth. At 99.9% light speed, the travel time for those on board decreases to a little over two months. At 99.99%, it's three weeks, and at 99.999%, a mere 7 days.

In principle, then, interstellar space travel is quite simple. By going fast enough, you can reach any part of the known universe in a human lifetime.

This sounds good, but it turns out that there's no such thing as a free launch. Relativity exacts a heavy price for time dilation. While perceived time slows down for high-speed space travelers, the mass and, consequently, the inertia of their rocket increases. More energy is required to produce acceleration; the gas pedal, if you will, starts to turn sluggish. This brutal fact makes travel at relativistic velocities an enormous, possibly insurmountable effort. No one yet knows how to build a rocket capable of safely traveling at 99% the speed of light or more.

But there might be other ways to reach the stars. One straightforward approach is to avoid the difficulties of high-speed cosmic cruising altogether. Accept the leisurely ride that low-tech rockets can manage, and build an "ark" commodious enough to house and comfort the many generations that will be born and die during the trip. The problems confronting interstellar travelers then become more sociological and less technical. Even so, they may pose a formidable challenge. Most readers have first-hand knowledge of the frictions that arise during the course of a one-week automobile trip. Imagine the damaging dustups likely to arise in a small colony of, say, a thousand folk confined to a rocket for a millennium. They might quickly fragment into warring camps, battling over mates or meals. The ark could reach its destination dead on arrival.

Even assuming that fatal feuding could be avoided, there are other dangers facing those that embark on an ark. Austrian sociologists Mircea Pfleiderer and Paul Leyhausen worry that the passengers might suffer a crippling loss of technological knowledge during their long flight, a real possibility in a small population. The original crew of the spaceship, which in short order becomes mere history, will be the last generation to actually experience the reality of living on a planet, with an environment filled with other living beings. Their descendants will inevitably lose touch. According to Pfleiderer and Leyhausen, "they will be culturally impoverished to stone-age level. They [will] have lived for too long in a closed, artificial 'universe' where perfect technical performance is mandatory..." Additionally, the limited gene pool would eventually lead to a crew with less ability to adapt. Highly inbred, the ship's passengers might fail when challenged by new situations or environments.

The manifold social problems of long trips could be solved in one of two ways: either place the passengers in suspended animation, thereby keeping them unconscious and out of trouble during their tedious voyage, or extend everyone's lifetimes so that the original crew actually reaches the ship's destination. Alas, despite the fact that both of these approaches are frequent staples of science fiction, no one knows how to do either.

Interstellar travel is not "on" for us, although we might manage to launch some small robot probes within the foreseeable future. But even these would not provide us with an answer to the question, "Are we alone?", within our lifetimes. At best we might hope that the probes will radio back some interesting information a few centuries hence.

The truth is simple: we won't find the aliens by blasting off in their direction. And the same goes for efforts to contact our cosmic brethren by gluing greeting cards to the sides of spacecraft—something we've occasionally done. In March, 1972 the Pioneer 10 probe was launched on its spectacularly successful voyage to the outer planets. Eleven years after blast off, it passed the orbits of Neptune and Pluto and became the first man-made object to leave the solar system. At the time of this writing, it is 6 billion miles from Earth, half again as far as Pluto. Attached to this distant, and largely inactive, hunk of space hardware is a 6 by 9 inch gold-anodized aluminum plaque, sporting an engraved pictogram conceived by Frank Drake, and Carl and Linda Sagan. The plaque is a message from Earth, and an admirable exercise in how to communicate something about ourselves to aliens—in this case, the putative critters who might someday retrieve Pioneer 10.

The engraving is both self-explanatory and informative, depicting our location in the Galaxy, Earth's place in the solar system, and a human couple stepping out of the shower to greet the aliens with upraised hand. An identical plaque was bolted to Pioneer 11, launched a year after its predecessor. The idea of attaching a one-page encyclopedia to probes headed for deep space clearly caught hold. In 1977, two more spacecraft were flung towards the outer solar system. Voyagers 1 and 2 sported gold-plated copper records containing over 100 photos of life on Earth, musical selections ranging from Chuck Berry to Bach, and spoken greetings in five dozen human languages and one so-called whale language.

Of course, you might wonder who gave NASA permission to send

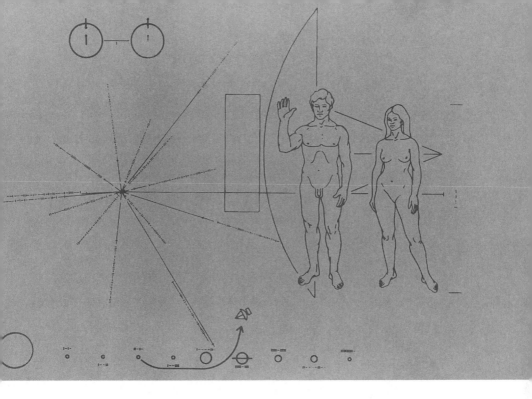

The plaque placed aboard Pioneer 10 is an ingenious attempt at universal communication. The radiating lines at left represent the positions of 14 pulsars arranged to indicate our Sun as the home star of the launching civilization. The "1–" symbols at the ends of the lines are binary numbers that represent the frequencies of these pulsars at the time Pioneer 10 was launched, relative to that of the hydrogen atom, shown at the upper left with a "1" unity symbol. The hydrogen atom is thus used as a universal clock, and the regular decrease in the frequencies of the pulsars will enable another civilization to determine the time that was elapsed since the launch of Pioneer 10. The hydrogen atom is also used as a universal yardstick for sizing the human figures and outline of the spacecraft shown on the right. The hydrogen wavelength—about 8 inches—multiplied by the binary number representing "8" shown next to the woman gives her height of 64 inches. Across the bottom are the planets, ranging outward from the Sun, with the spacecraft's trajectory arcing away from Earth, passing Mars, and swinging by Jupiter. The human figures, of course, represent the creators of the plaque and Pioneer, with the man's hand raised in a gesture of good will. [NASA]

maps of our home into space. Could this be a dangerous move? Would aggressive aliens, alerted to our presence and location by these delicate engravings, swoop down on Earth to steal our molybdenum or abduct a few primates for breeding purposes? Well, no. At seven miles per second, the speed with which these probes are distancing themselves from the Sun, it will be 70,000 years before they reach the nearest star. And, of course, they're not *aimed* at the nearest star. They will randomly enter another solar system about once per billion years, by which point the consequences to those who financed and launched the probes will be minimal.

Not only is there no danger from sending informative graphics into space, there's also very little chance that they will ever precipitate contact with extraterrestrials. You might compare these efforts to a scenario that might have taken place in 1492. Isabella and Ferdinand, the ruling monarchs of Spain, decide that they are unable to grant Columbus the monies needed to build the Nina, Pinta, and Santa Maria, despite good reviews by the referees on his proposed project. Instead, they send him to the beach with a bag full of glass bottles, suggesting that he put notes in them beseeching Indians to "please reply." This method of discovery, like the plaques and records on our spacecraft, is painfully slow and enormously uncertain.

Despite the inefficiency of interstellar plaques, we have to admit their romantic appeal. They are the ultimate time capsules. The informative license plates on the Pioneer craft and the "Earth's Greatest Hits" gold records on Voyagers 1 and 2 will still be readable many millions of years hence. They stand a good chance of outlasting humanity's dominion on this planet. But unlike earthly time capsules, the odds are strongly against their ever being read again.

BROADCASTING MESSAGES

Sending hardware-borne inquiries into space takes too long. But light and radio waves can bridge the vast interstellar deeps at 186,000 miles per second. Could we not prod the aliens into revealing their presence by beaming them a message—a hailing signal that would reach them in years, not eons?

In some sense we've been unintentionally doing this for most of the 20th century. Light and radio from our cities have been leaking willy-nilly into space. Some of the harsh glow of street lamps spills into the night sky, and is easily seen by satellites orbiting Earth. Such stray illumination, sometimes called "light pollution," is a serious annoyance to astronomers on our planet. But it would be thoroughly invisible to alien stargazers. The nighttime brilliance of even major metropolitan areas such as New York or Tokyo is enormously weakened by distance, and hopelessly confounded by our Sun's thunderous glare. The aliens won't see it.

At radio wavelengths, the chances of detecting earthly activities are less bleak. To begin with, the interference by the Sun is not as troublesome because Sol funnels most of its energy output into light, not radio. In addition, transmitters on Earth usually concentrate their energy in preferred directions, and confine it to a narrow part of the radio dial. The result is that at many radio frequencies, the Earth is far and away the "brightest" object in the solar system, possibly detectable by aliens outfitted with hi-tech scanners and big antennas.

A word about radio. When we speak of radio, we're talking about light of long wavelength (or if you prefer, of low frequency). Radio and light are both electromagnetic radiation, and travel through space at the already-noted speed of 186,000 miles per second. A radio signal beamed from Earth can reach the moon in a little more than a second, Mars in five minutes, and Alpha Centauri in 4.5 years. Wavelengths normally defined as "radio" range from hundreds of meters (low-frequency radiation mostly used for clandestine communications with underwater submarines) to ultra-high frequency millimeter waves (used for some radars and satellite communications).

While all radio wavelengths have proven useful on Earth (indeed, so useful that there is continual international bickering over who can use which wavelengths for what), certain parts of the dial are more suitable for getting in touch with E.T. than others. The relatively low-frequency signals that typified radio development in the early part of this century, and which are still used for AM broadcasts, reflect easily off the Earth's ionosphere, a blanket of charged particles high above our heads. If your interest is in picking up a distant Top 40 AM station on your car radio, this ionospheric reflection is a good thing. Your favorite clear-channel station in Chicago will be reflected down to your car in Florida. However, the ionosphere ensures that very little of the energy in an AM broadcast ever makes it into space. The aliens have no chance of hearing such stations. Perhaps they should be grateful.

Television and FM radio signals lie at frequencies about 100 times higher on the dial than AM. These high-frequency broadcasts are *not* reflected by the ionosphere, a fact that serves a useful earthly purpose. Since television, in particular, takes a lot of bandwidth (or, if you prefer, space on the dial), the number of channels that can be parceled out to TV stations is limited. Since broadcasts from New York won't reflect back down on, say, Detroit, the government can assign the same channel to stations in both cities. Television signals don't reach potential viewers beyond the horizon. The signals pass through the ionosphere, and off into the universe, where alien audiences may be waiting.

It is now more than a half-century since large-scale television broadcasting got underway. Those early programs are premiering tonight 50 light-years away in the depths of space. Several thousand star systems have been bathed in the faint glow of Earth's TV transmissions. If any of them host inhabited planets, it's possible that sophisticated extraterrestrials have detected these signals. Perhaps alien fans of *Mr. Ed* are trying to puzzle out whether it's the two-legged or the four-legged creatures who rule Earth.

Our television signals could be detected, assuming that the aliens are close enough for the broadcasts to have reached them. But cosmic couch potatoes would require enormous antenna farms to receive these early broadcasts. At 50 light-years' distance, the aliens would need to cover 3,000 acres with rooftop-style TV antennas in order to pick up a typical big-city television broadcast from Earth. That's just to detect the signal, to know that there's a transmitter on the air, not to pick up the "message" (the picture and sound). To actually see Mr. Ed shuffle about his stable or listen to his philosophical whinnies, the extraterrestrials would need to construct an antenna array tens of thousands of times larger.

High-powered military radars, as well as radars used for mapping of nearby planets, have sent far stronger signals into space. Of course, radar is intermittent and highly directional, while our television broadcasting goes on day and night. However, if an alien civilization happens to be listening at the right time, and is located in the beam of one of these specialized transmissions, it could be the recipient of a clearly artificial, if rather uninteresting signal.

Although Earth has been betraying its presence for half a century, deliberate hailing signals have generally not been sent to putative alien audiences. Most astronomers, including those who actively search for aliens, have not favored an active effort to get the attention of the extraterrestrials with a radio inquiry. The reason is simple. The nearest alien society might be 100 or 1,000 light-years away. A message would take centuries to get there, and if the extraterrestrials both exist and deign to respond, it will take a similar amount of time for their pithy reply to reach us. After centuries of waiting, the earthly researchers who conceived the project would undoubtedly have lost both interest and funding.

Nonetheless, on one celebrated occasion, a deliberate transmission of high power was sent spaceward, intended for alien ears. On November 16, 1974, as part of the dedication ceremonies for an upgrade of the Arecibo Radio Telescope in Puerto Rico, a simple picture was broadcast in the direction of the globular star cluster M13. Arecibo is the world's largest radio telescope, and sports a metal reflector 1,000 feet in diameter. It is fitted with a transmitter for radar studies of the upper atmosphere and the planets. When the transmitter is fired up, the huge

The Arecibo radio telescope, Puerto Rico.

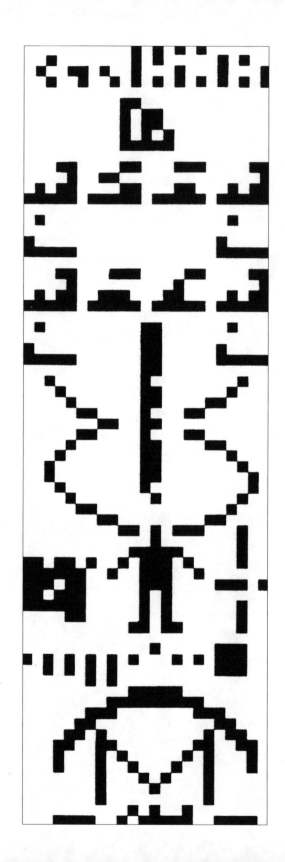

reflector ensures that the energy is focused into a tight beam (in the case of the 1974 transmission, the beam was about 2 arcminutes across. If the beam were aimed at the moon, the illuminated spot would be about the size of one of the larger craters.) The Arecibo transmitter was switched back and forth between two very close frequencies, sending a sequence of 1s and 0s during the three minute transmission. If any recipients are clever enough to arrange these 1s and 0s on a grid, they will be rewarded by a simple graphic depicting the telescope, a few facts about our solar system and our biochemistry, as well as a discreetly sexless human figure.

The artwork broadcast during this exercise is reminiscent of the Pioneer 10 plaque shot into space two years earlier. However, there are noteworthy differences. The Arecibo message was sent in the direction of known stars—indeed, M13 is a conglomeration of several hundred thousand stars. Admittedly, it's a long way off, approximately 21,000 light-years. Consequently, it will take 21,000 years for the Arecibo signal to reach this rich stellar neighborhood. But even after this prodigious period of time, the Pioneer craft will still not have encountered its *first* star. Additionally, the Pioneers are small, dark objects the size of an automobile, and would be exceedingly difficult to find even if they should eventually cruise within a light-year or two of some alien's home planet. Finding the Arecibo broadcast, even at M13's distance, would be straightforward, requiring only a very large antenna and a bit of luck. (The recipients need to have their antenna pointed in the right direction and be tuned to the correct frequency during the relevant 3 minutes.)

If your mission is to wake up the aliens, sending bits rather than rocket-powered bric-a-brac is the way to go. It is both enormously speedier and likely to reach a larger potential audience. Indeed, this point was not lost on Britain's Astronomer Royal, Sir Martin Ryle. Shortly after the Arecibo broadcast, Ryle petitioned the International Astronomical Union to rein in astronomers eager to get in touch with cosmic neighbors. Obviously, any alien recipients of the graphic message sent from Puerto Rico might be able to pinpoint its planetary source. Ryle's con-

The Arecibo radio message.

149

cern was that this might be dangerous, a possibility we have already considered. Few astronomers seem to agree, especially as the Arecibo transmission was merely a 3-minute demonstration, not a serious effort to establish communication with the aliens. And as several researchers were quick to point out, the Astronomer Royal had made no efforts to shut down either BBC Television or the world's radars.

RADIO SEARCHES

Passive radio searches, collectively known by the acronym SETI (Search for Extraterrestrial Intelligence), have become the preferred approach for tracking down thinking beings in space. We don't transmit. We listen, hoping to tune in on signals that may have left their native worlds long ago.

The idea of eavesdropping is not new, and modern efforts had some early and interesting precedents. Just as the 20th century was dawning, the electrical pioneer Nicola Tesla tried to attract the attention of nearby extraterrestrials with long-wavelength radio static. The optimistic Tesla was even of the opinion that his noisy calls had been answered, and he assumed that the signals he heard one night came from Mars. In fact, they were undoubtedly "whistlers," radio waves generated in the atmosphere by lightning and other disturbances.

In 1924, there was a special opportunity to listen in on planetary party lines from everyone's favorite alien world, Mars. The red planet was in a particularly favorable opposition, which is to say it was as close to Earth as it ever comes, roughly 36 million miles. It was during a similar opposition in 1877 that Schiaparelli first spotted his *canali,* and although a half-century had passed, the idea of intelligent life on Mars was still very much alive. The new technologies of radio offered a way of possibly proving the Martians' existence: we could tune in their news broadcasts. For three days in August, U.S. Navy radio transmitters across the Pacific were voluntarily silenced in order to facilitate listening for the expected transmissions. This was a front page story in the *New York Times,* which noted that the crackerjack cryptologist William F. Friedman, chief of the U.S. Army's code section, was standing by to unravel any messages our alien neighbors might be sending our way. Needless to say, no signals from Mars were heard. Similar listening attempts by radio pioneer Guglielmo Marconi several years earlier had also failed to uncover red-planet radio traffic, although a Seattle-based transmitter caused Marconi some momentary excitement.

These efforts seem naive in retrospect, mostly because of what we now know about Mars. Given the fact that radio astronomy was unimaginable in the 1920s, these simple experiments were remarkably far-sighted.

Science marched on, and by the late 1950s, the space age had dawned. While many researchers still thought that life on Mars was likely, few believed it was sophisticated enough to build radio transmitters. Intelligent aliens, if they existed at all, would be at great distances. Consequently, some scientists began to consider radio as a means of communicating, not across the solar system, but across the Galaxy.

The seminal work on this idea was from two Cornell University physicists, Giuseppe Cocconi and Philip Morrison. In 1959, Cocconi was studying gamma rays produced by the Cornell synchrotron. It occurred to him that a narrow beam of these high-energy rays might be a practical signaling device in interstellar space. When he told Morrison his idea, the latter suggested that gamma rays were maybe not the best way to communicate. Radiation at longer wavelengths— radio waves—might be more efficient. After making a simple calculation, the two physicists found, somewhat to their surprise, that existing antennas on Earth were capable of sending a radar signal detectable at a distance nearly that of Alpha Centauri. This was already impressive, but extraterrestrials for whom radio was an old technology could presumably do far better.

Radio waves, easily and inexpensively generated, could reach across the galactic voids. Advanced civilizations might long ago have filled the interstellar ether with radio traffic. If so, we could learn of their presence simply by listening. Cocconi and Morrison urged radio astronomers, who had the large receiving antennas necessary to make a sensitive search, to examine neighboring, sun-like stars. They even told the astronomers where on the dial to tune. Cocconi and Morrison argued that the best frequency to use for star-to-star communication was 1420 MHz. For more than a dozen years, radio astronomers had been turning their receivers to 1420 MHz (21 cm wavelength) to study the tenuous, hydrogen gas that floats between the stars. The gas, used as a tracer for mapping the extent and motions of our galaxy, produces natural cosmic static at 1420 MHz. It is the most important frequency in all of radio astronomy, a fact that advanced aliens will appreciate as well. It is a "universal channel." Anyone wanting to be heard would use it as a hailing frequency.

Within a few months, Cocconi and Morrison had published their ideas in the research magazine *Nature*. While the popular press found the idea of listening for extraterrestrials exciting, many astronomers reacted skeptically or even critically. But the two Cornell physicists were aware of the profound importance of a successful search. They countered skepticism by urging that their ideas be put to the test, that an experiment be done. The *Nature* article concluded with a challenge to those who would consider such an experiment too much of a long shot:

"We . . . feel that a discriminating search for signals deserves a considerable effort. The probability of success is difficult to estimate; but if we never search, the chance of success is zero."

Cocconi and Morrison didn't know it, but a young astronomer would soon take up their challenge. The modern search for cosmic company was about to begin.

SETI

In the spring of 1960, Frank Drake was 29 years old, full of juice, and unknown. Within a few months, his anonymity would be lost forever. He was about to become the father of SETI.

The ink on his Harvard Ph.D. was still wet when Drake accepted a staff position at the National Radio Astronomy Observatory (NRAO) in Green Bank, West Virginia. The Observatory was, like Drake, young. In 1960, its only real instrument was an 85-foot radio telescope, a modest device by today's standards. The small antenna would be the first radio ear to be cupped for the detection of alien signals.

The NRAO sprawls across a mile-wide valley between wooded hills that abut the main spine of the Alleghenies. The seclusion is deliberate. West Virginia's mountains work as natural barriers against radio interference, and the low population density ensures that there isn't much man-made static to begin with. In the 1970s, the song writer John Denver famously described the mountain state as "almost heaven." To those who eke out unimpressive paychecks in the hardscrabble towns of the hollows, Denver's description is just wishful thinking. But there is no denying that West Virginia, if not heaven itself, is a good place for *studying* the heavens, at least at radio wavelengths.

Nearly all newly minted Ph.D.s begin their research careers with experiments that are souped-up repeats of their doctoral theses. Not Drake.

He chose a highly unconventional project to kick off his tenure in Green Bank. Rather than study the natural radio noisemakers of the universe—the mammoth clots of gas that litter the Milky Way, or the electrically charged innards of distant galaxies—Drake elected to search for artificial broadcasts, signals deliberately beamed by an alien society.

"I was crazy. Since childhood I had an interest in whether there were intelligent creatures on other worlds," says Drake. "But suddenly I was in a position to do something about answering that question."

He decided to listen for signals near 1420 MHz, the well-known hydrogen frequency advocated by Cocconi and Morrison. The two MIT physicists had reasoned that advanced aliens would surely know of the steady hiss made by interstellar hydrogen. If extraterrestrials wished to get the attention of unknown beings light-years away, they would tune their radio transmitters to frequencies near 1420 MHz, using the hydrogen as a 'marker'—a signpost that says "listen here." Drake, who hadn't read these arguments, nonetheless chose the same frequency band. His reasoning was more pragmatic. Receivers for studying hydrogen already existed at the NRAO, and by using them, he said, "I avoided any criticism that I was building specialized, costly equipment for one-time use in a wacky experiment."

Because it was unclear how closely E.T.'s signal might hug the hydrogen channel, Drake decided to scan a limited range (400 kHz) of frequencies centered at 1420 MHz. Rather than twiddle the knob himself, Drake rigged a small motor to mechanically tune the receiver up and down the dial. For monitoring the receiver's output, he used a chart recorder and, as dramatic complement, a loudspeaker. The total cost of the equipment was about $2,000.

Drake planned to spend a couple of months hunting for signals. His prey were two dim stars in the southern sky, Tau Ceti and Epsilon Eridani. Although inconspicuous, these stars are special. They are solitary sun-like stellar systems, just under a dozen light-years distant. Indeed, Tau Ceti and Epsilon Eridani are the nearest stars about which we might reasonably expect earth-like planets to spin.

The idea was exciting, although the execution was sometimes uncom-

Frank Drake, Director of The SETI Institute.

fortable. "The weather in April was cold, and I was working alone in the small control building for the 85-foot. Every morning, first thing, I had to crawl into the telescope's parametric amplifier to tune it. That was really unpleasant," recalls Drake. "I ran the telescope to Tau Ceti, and found only the familiar, noisy background hiss of, well, nothing special. I listened for about a half day to that, and then the star set— dropped below the horizon."

"When Tau Ceti was gone, I simply slewed the antenna to the other target, Epsilon Eridani. A few minutes went by, and at first there was only a waterfall of noise, like before. But shortly, and quite unexpectedly, there came this horrendously loud 'whoosh, whoosh, whoosh' over the loudspeaker, about eight times a second. I was incredulous, and sat there momentarily frozen. My first thought was, 'Could it be this easy?' My second thought was, 'What do we do now?'"

Drake hadn't expected such an early success for SETI. The lack of foresight didn't matter. The success was illusory, as his signal soon went away. Ten days later, it briefly returned, but this time the excitement was muted. The signal was also being picked up by a small antenna aimed at a completely different part of the sky. Drake eventually concluded that the noisy excitement was due to radar from a high-altitude military plane. It was evidence for military intelligence, not extraterrestrial intelligence. No convincing signals were heard from either star. The aliens would not show themselves the first time out.

Despite the lack of an immediate discovery, Frank Drake's pioneering experiment ignited strong interest from both the science community and the public. Drake had even given his experiment a name: Project Ozma. Ozma, of course, was the princess of the land of Oz in Frank Baum's books. "Oz was a land far away, difficult, and populated by strange and exotic creatures," Drake said. But the use of a name was a sublime move. Although telescopes are routinely baptized with bluntly descriptive monikers such as "the 140-foot telescope" (known to Europeans as "the 42-meter telescope"), observing programs are seldom christened. By giving his experiment a name, Drake had given it a face. Subsequent SETI searches would copy his example.

This ground-breaking effort to eavesdrop on E.T. encouraged dozens of imitations. Many of these were *ad hoc* efforts by radio astronomers who found themselves with a few extra hours of radio telescope time on their hands. While waiting for the next galaxy or pulsar to rise above the horizon, some observers would swing the antennas in the directions of nearby stars or other suspected alien habitats, hoping for the big payoff. These casual mini-Ozmas were followed by more deliberate attempts to improve on Drake's experiment, either by scrutinizing a larger number of star systems or by opting for a different sort of target altogether. For example, Boston astronomer Nathan Cohen and his colleagues sought signals from stellar clusters. Instead of checking out possible alien star systems one at a time, Cohen et al. examined an entire stellar community at once. Such clusters contain several hundred thousand stars, an impressively large number of potential alien homesteads. Since they are born together, cluster stars are all nearly the same age. Consequently, if civilizations exist in these crowded starfields, they might be similar enough in technological level to establish communication links—to engage in interstellar signaling that we might chance to overhear.

Other experiments have sought E.T.'s radio whine in the direction of the Galaxy's only cartographically unique location, its center. The Milky Way's inner sanctum is not a placid place; the thick starfields of the center are rent by intense bursts of radiation, and a massive black hole probably hulks there as well. Despite the dangers of these central realms, a truly advanced civilization might site a radio beacon there on the reasonable assumption that, sooner or later, every other galactic society will aim their radio telescopes at the Milky Way's downtown districts.

Still other radio astronomers have pointed their telescopes towards neighboring galaxies. By so doing, they can eavesdrop on billions of stars at once. Of course, even the nearest galaxy (the Large Magellanic Cloud) is 150,000 light-years away, or thousands of times farther than the stellar targets of Project Ozma. E.T.'s transmitters would need to be extraordinarily energetic to be heard from such stupefying distances. But astronomers reasoned that in an entire galaxy there may be a few sophisticated societies in command of such powerful technology.

None of these early projects managed to turn up a single, convincing extraterrestrial signal. By the 1970s, astronomers who took SETI seriously concluded that informal searches were unlikely to find cosmic company. To begin with, they felt that a hunt should consist of more than occasional forays to flush E.T. from a handful of habitats. A prolonged, systematic search was required. After a decade of small-scale experiments, it was becoming clear that the sky was not teeming with strong, unmistakable extraterrestrial signals. Finding E.T. was not going to be easy.

Another thing that became apparent was that to do SETI right would require some decent tools. SETI scientists use the same, large telescopes that are routinely trained on galaxies, quasars, pulsars, and other loud members of the astronomical bestiary. But the type of signals they seek differs from natural signals. A quasar, for example, pumps out an enormous amount of radio energy. But this energy is spread all over the dial. If you tune the knob of a receiver that is listening to a quasar, you'll hear static (the radio energy produced by the quasar) across most of the band.

A signal from intelligent beings, on the other hand, would probably be confined to a small part of the dial. Narrow-band broadcasts provide more "bang for the buck" if your intention is to transmit a signal easily heard at great distances. The energy of the radio transmitter is concentrated in a narrow range of frequencies, much the way the screech of a locomotive whistle is confined to a small range of audio frequencies, or tones. It's only reasonable to assume that signals produced by alien transmitters will be at least partly "narrow band." To find them, SETI researchers want receivers that can break up the incoming cosmic static collected by their radio telescopes into narrow channels. Clearly, this is rather different from the equipment needed to plumb the secrets of quasars or other natural phenomena. The receivers that early SETI researchers borrowed from radio astronomers were great for studying galaxies, but bad for finding aliens. They reduced sensitivity to any artificial, narrow-band signals by factors of hundreds to thousands. It was as if explorers were forced to wear sunglasses while searching for cats in the dark.

Narrow band receivers that slice and dice the radio dial are the answer. And building such a receiver isn't particularly hard. Drake did it for Project Ozma; his equipment had a single channel that was swept up and down the band by a small motor. But sweeping a single channel, scanner-like, isn't an efficient way to search for alien signals either. The reason boils down to the fact that E.T. has never sent a fax or e-mail telling astronomers exactly where on the dial his transmissions can be found. Cocconi and Morrison did their best to anticipate alien broadcast strategy by suggesting that they would transmit at microwave frequencies near the 1,420 MHz neutral hydrogen line. Many modern researchers agree. After all, not only is there a clear-cut 'marker' here, but this part of the band suffers the least interference from natural, cosmic static.

Still, there are billions and billions of channels to search, even restricting oneself to microwaves (Project Ozma only examined a tiny fraction of these). Listening to a few billion channels one at a time would exasperate Job. Imagine scanning through the band, stopping for a minute at each frequency to check it for E.T.'s signal. It would take close to two thousand years to get through even the first billion channels, somewhat longer than the duration of the average research grant. So scanning for alien signals is impractical. SETI researchers quickly realized that they needed custom-built receivers that could monitor many millions of channels *simultaneously.*

The outfitting required for a serious extraterrestrial hunt includes multichannel, narrow-band receivers, large radio telescopes, and a systematic search program. In the early 1970s, these requirements were quantified in a comprehensive NASA report known as "Project Cyclops," a hulking document reverently known to insiders as SETI's bible. Based on a summer study initiated by John Billingham, a former British military doctor working at the Ames Research Center in California, Cyclops distilled the expertise of astronomers and engineers from both NASA and academia. The report was largely written by Barney Oliver, Vice President for Research and Development at the Hewlett-Packard Corporation. Oliver, a man of razor-sharp intellect, a booming voice and a looming presence, had been fascinated by the challenge of SETI since its beginning. He had visited Drake in Green Bank during Project Ozma and grasped both its importance and its technical limitations. Project Cyclops envisioned constructing an array of antennas whose sole purpose would be to scan the heavens for extraterrestrial signals. Beginning with a handful of telescopes, the array would be expanded until success was booked.

The Cyclops array was never built, but the technical analysis embodied in the report was visionary, and is still germane today. Of equal consequence, Billingham and Oliver's landmark study prompted NASA involvement with SETI, a stormy love affair that lasted two decades.

The NASA SETI program began small and grew to be modest. Its strategy for uncovering cosmic companions was far less ambitious than that proposed by Project Cyclops, but nonetheless borrowed heavily from that landmark study. The approach was two-sided. One side was to use the antennas of NASA's Deep Space Network in a systematic scan of the whole sky. The other, which would employ the world's largest radio dishes, was a targeted scrutiny, *á la* Project Ozma, of about one thousand nearby, sun-like star systems. The sky survey component was run by scientists at the Jet Propulsion Laboratory in

Pasadena, California, while the targeted search was based at Ames, in Mountain View, California. The latter center, under Billingham's leadership, was in charge.

In its early years, funding for this program was a few million dollars per year, or roughly two cents annually per taxpayer. Despite the small budget, some congressmen fretted that citizens might not be getting their two cents' worth. In 1981, Senator William Proxmire expressed his scorn for NASA SETI by giving it his 'golden fleece' award and deleting its funding. Alarmed astronomers, including science icon Carl Sagan, managed to turn the stingy senator around, bringing the program back into the budget a year later. But by 1992, with the first observations just getting underway, the program had grown to a modest $12 million annually, raising the taxpayer burden to nearly a dime a year. Congressmen eager to appear effective at deficit reduction were once more looking critically at SETI. Sensing danger, a staff worker for Senator Barbara Mikulski, who was then chairman of the Senate appropriations committee governing NASA and a supporter of the SETI program, suggested a pre-emptive move: change the project's name to the dull-sounding 'High Resolution Microwave Survey.' Apparently the hope was that by cloaking the effort in bureaucratese, critical legislators might leave NASA's small but controversial back-burner program alone.

SETI astronomers chafed at this unsubtle obfuscation, but had no choice but to go along. Survival, after all, is everything. The program was soon referred to (if not widely known as) the HRMS. There was no hiding behind nomenclature, however. In October, 1993, Senator Richard Bryan of Nevada introduced an amendment to kill NASA SETI. Not reduce it by a painful 10% or cripple it with a 50% cut. Bryan's amendment zeroed out SETI entirely. The amendment passed. In arguing the wisdom of this move, the senator's press release crowed that the NASA program had "failed to bag a single little green fellow."

At the time of its demise, fewer than 0.1% of the proposed observations for the HRMS had been completed. It was as if Isabella and Ferdinand had spent the big pesos and more than a dozen years to build three ships for Columbus, but had suddenly canceled his voyage while the explorer was still in sight of the Spanish docks. Senator Bryan saved American citizens a few cents each per year, but had brought to a halt the most ambitious search ever begun to find thinking counterparts elsewhere in the Galaxy. Somewhat ironically, two years later a platoon of Nevada politicians attended ceremonies to grandly announce the opening of an 'extraterrestrial highway,' a hundred-mile stretch of state route 375 that passes near the secret military facility known as Area 51. This Air Force test site is widely and irrationally believed to be a clandestine storage depot for crashed aliens and their high-flying hardware. UFOs and government cover-up were more in keeping with the mindset of Nevada politics than a scientific search for extraterrestrials. Alien asphalt, after all, had an immediate payoff in tourist dollars.

SETI, always vulnerable, had once again suffered a mortal wound. Federal funding was gone, and revival of congressional support was not in the cards given the strong political pressures in 1993 to trim spending. It looked as if the first generation of humans able to discover its role in the drama of the universe wasn't going to bother trying.

However, a peculiar bureaucratic circumstance came to SETI's rescue. For nine years the Targeted Search portion of the NASA program had been administered by the non-profit SETI Institute, a small organization situated in Mountain View, a few miles from NASA's Ames Research Center. The original purpose of the Institute, as promoted by its founder and Executive Director, Tom Pierson, was to better use monies allocated to SETI by minimizing the associated bureaucracy. Government research programs funded by NASA or the National Science Foundation are typically compelled to spend one-third or more of their dollars on administrative overhead, office space, etc. The SETI Institute's efficient, private management of these support services cut such overhead by half.

But what had been established for efficiency now became key to survival. As a non-government organization, the SETI Institute could, in principle, pursue the search with private money, irrespective of the ebb and flow of congressional sentiment. The project, or at least some of it, could continue if the money could be found. At this point, several key people associated with SETI were working with or at the Institute, and were able to press the effort to find a life jacket to keep the search afloat. The Institute president was Frank Drake, SETI's paterfamilias and an astronomy professor at the University of California, Santa Cruz. Jill Tarter, a long-time SETI researcher and a real-life analog to Ellie Arroway, the radio astronomer of Carl Sagan's 1985 novel (and 1997 film) *Contact,* headed up the Institute's search efforts. Cyclops author Barney Oliver, who, after his retirement from Hewlett-Packard Corporation, had been Deputy Chief of NASA SETI under Billingham, immediately gave up his part-time Civil Service position to help the SETI Institute. Oliver soon had his former employers, Bill Hewlett and Dave Packard, on the phone. Within weeks, these two high-tech entrepreneurs, as well as Gordon Moore (co-founder of Intel Corporation) and Paul Allen (co-founder of Microsoft) pledged a million dollars each from their private bank accounts.

The NASA SETI program was dead, but a new, privately funded project modeled on the targeted search component of the HRMS would continue. Barney Oliver urged naming the reincarnated, privatized search "Project Phoenix." Despite some muted opposition (including a concern by me that the name sounded like a condo development in the American southwest), the Institute liked Oliver's allusion to the mythological bird that rose from the ashes. Project Phoenix was born, and in February, 1995 resumed the systematic scrutiny of nearby, solar-type star systems that had been scrapped two years earlier.

A HANDFUL OF PROJECTS

With government funding gone, Project Phoenix became the most ambitious SETI experiment underway anywhere. Its strategy was a direct descendent of Drake's Project Ozma: pick out nearby, sun-like stars, and listen for narrow-band, microwave radio transmissions coming from their directions. Since E.T.'s signals are expected to be weak, Phoenix uses the largest antennas it can muster for its cosmic eavesdropping. It began its search using the 210-foot Parkes radio telescope,

The 140-foot radio telescope at Green Bank, current site of Project Phoenix.

situated in sheep country about 250 miles west of Sydney, Australia. In 1996, it continued its study of a thousand, close-by solar cousins using the 140-foot radio telescope in Green Bank, West Virginia. The latter instrument is but a short walk from the small antenna used by Frank Drake for Project Ozma nearly four decades earlier.

Phoenix examines a total of two billion channels for each of its stellar targets, covering the microwave band between 1,000 and 3,000 MHz. It does so by listening in a block of 28 million simultaneous channels for five or ten minutes, then shifting the block up the radio dial and listening to another set of 28 million channels. It takes about a day's worth of listening and shifting to search one star over the two billion channels that the Phoenix astronomers think are the best bets for alien broadcasts.

It may come as a surprise, although it shouldn't, that this system finds signals all the time. After all, Phoenix (and indeed, most SETI projects) use enormous antennas plugged into sensitive, multichannel receivers. Narrow-band signals, most of them from radar or telecom-

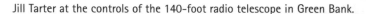

Jill Tarter at the controls of the 140-foot radio telescope in Green Bank.

munications equipment, show up routinely. The growing hordes of satellites that pirouette above our heads have saturated the airwaves with signals that, at first blush, often appear to be alien transmissions. Project Phoenix has tried to inoculate itself against this noisome contamination by always using two radio telescopes simultaneously. When a signal is found with the primary telescope, its characteristics are immediately sent to a second antenna, situated hundreds or thousands of miles away. Thanks to the Earth's rotation, the signal frequency from any extraterrestrial transmitter should be ever-so-slightly different when received by the two antennas, a small shift on the radio dial that, in fact, can be easily measured. Phoenix uses this minor de-tuning to sort transmission wheat from chaff. The overwhelming majority of the confusing, man-made interference picked up by the Phoenix system is recognized (and then rejected) within minutes by this tandem-telescope scheme.

Fortunately for the motivational needs of the astronomers involved, some signals survive the automated weeding process and cause a bit of excitement. Several times a week, a candidate alien signal will be flagged as "looking good" by the software that monitors the receivers. One channel in the sea of 28 million will bear the hallmarks of extraterrestrial origin, and will have passed the highly-efficient two-telescope test described above. In that case, the control software will swing the telescope a few degrees away and look again, pointing not at the targeted star system but at blank sky. If the signal disappears, as would be expected if it came from the system under scrutiny, the telescope is aimed once more at its target. A true signal from E.T. would then reappear. By repeating these "off and on" tests, the astronomer can quickly ascertain whether the signal is simply interference from an earthly satellite, or whether it warrants heightened blood pressure and visions of award ceremonies in Stockholm. No signal so far has passed both the two-telescope test and repeated off-and-on scrutiny.

Phoenix may be the biggest kid on the SETI block, but it's certainly not the only one. The longest continuously operating SETI experiment is the Big Ear project at Ohio State University. All day, every day, the Ohio State telescope scans the skies for E.T.'s chatter. Alas, a decision to scrap the telescope means that this experiment, begun in 1973,

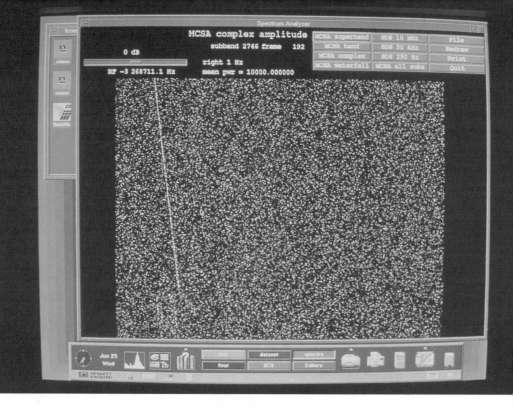

A display screen from the Project Phoenix SETI search. Spectral scans of approximately 600 of Phoenix's 28 million channels are seen running from left to right on the screen, with bright dots indicating more incoming energy than dark dots. Once per second, a new scan is added to the top of the screen, pushing off an old scan at the bottom. The diagonal white line is an extraterrestrial signal, although it comes from a man-made source: the Pioneer 10 spacecraft. The craft is approximately 6 billion miles distant, or 10 times as far as Pluto, and the transmitter power has dropped to about 5 watts. The ability of the Phoenix receivers to detect this dim, distant signal gives some idea of their enormous sensitivity. The signal is seen to be slowly drifting to lower frequencies. This is due to the rotation of the Earth, which introduces a changing Doppler shift.

will end sometime in 1998. University officials, making a strong statement about priorities, have chosen to allow developers to convert the observatory to a golf course.

The doomed Ohio State project is well known for a powerful signal it picked up in 1977. Indeed, the signal on the chart recorder paper so impressed Jerry Ehman, the astronomer on duty, that he wrote "wow" next to it. The so-called "wow signal" has become a staple of both science fiction and internet chat groups that believe that this clear evidence for extraterrestrials is being ignored. However, the "wow signal" was never seen again, despite searches at the same position and same frequency. Bob Dixon, the principal SETI scientist on the spot, strongly doubts that it was E.T. calling Ohio. Any extraterrestrial signal worth its salt, or at least worth mention in the serious news media, must be persistent enough to survive repeated efforts to confirm it. That's the way of science. If a neighbor claims to have seen a ghost in his attic, you might be tempted to camp out under the rafters with a video camera. But if you do so and never see the ghost, there are only two conclusions you can draw: (1) a ghost exists, visited your neighbor once, and then retired to Florida or elsewhere, or (2) your neighbor may have seen something, but it wasn't a ghost. If, on the other hand, you *do* see the ghost, and other neighbors can do the same, then you have good evidence that spooky beings are under your neighbor's roof. Newton wouldn't have bothered concocting laws of motion if a dropped apple was an experiment that could be performed only one time. Similarly, SETI scientists will only believe that a signal comes from E.T. if they can see it again and again. The "wow signal," regrettably, only showed up once. For this reason, it has failed to wow the scientific community.

Another SETI experiment is sponsored by the Planetary Society, a membership organization founded by Carl Sagan and Bruce Murray, a former Director of the Jet Propulsion Lab. The Planetary Society's SETI search, like the Ohio State project, is scanning large chunks of the sky rather than zeroing in on suspected alien abodes. Dubbed Project BETA (for Billion channel ExtraTerrestrial Assay), the Society's experiment uses an 85-foot telescope outside of Boston, Massachusetts connected to a 250-million-channel receiver. (The receiver's designer, Harvard electronics whiz Paul Horowitz, admits that 250 million is not quite a billion, but claims that the exaggeration serves to emphasize the dramatic improvement upon META, the Society's previous effort, which used an 8-million-channel receiver. The META system is now operating in Argentina, where it is being used to scan the southern sky.)

SERENDIP is yet another major American microwave SETI experiment. The acronym, concocted by Jill Tarter, stands for Search for Extraterrestrial Radio Emissions from Nearby Developed Intelligent Populations, a fact that the reader may have found obvious. Tortured acronyms litter the SETI landscape, incidentally. SETI searches have been dubbed SIGNAL, META, BETA, SETA, AMSETI, MANIA, and OZPA. An ingenious acronym apparently conveys a sense of ingenious science.

SERENDIP is piloted by Stu Bowyer and Dan Werthimer at the University of California, Berkeley. The experiment rides piggyback on radio astronomical observations made at the imposing 1,000-foot Arecibo Radio Telescope in Puerto Rico. By taking advantage of receivers *not* being used by astronomers, the SERENDIP crew can get 24-hour-a-day use of this, the world's largest antenna. Of course, there's a slight penalty to be paid. The Berkeley SETI astronomers have no say over what patch of sky they're pointed at, and can't even choose the radio band they'll be listening to. Nonetheless, over the course of two years, their 168-million-channel search randomly probes almost one-third of the total sky, looking for a serendipitous signal.

An amateur SETI effort also exists. Electrical engineer Paul Shuch heads an organization known as the SETI League whose goal is to enlist the help of radio amateurs in scanning the skies for intermittent, strong broadcasts. Although the sensitivity of the amateur equipment is very low, this strategy could work if the aliens beam a very powerful signal for a relatively short period of time, say days or weeks. Such a signal might escape the notice of the professional SETI searches.

Although American experiments dominate SETI activity today, other countries are beginning to prick up their metal ears. The Argentinian counterpart to the Planetary Society's sky survey has been mentioned. At the University of Western Sydney, Macarthur, another piggyback SETI search of the Galaxy, begun in 1998, is underway. The SETI Australia project will use the 210-foot Parkes telescope to feed a receiver system based on the one built by Dan Werthimer of the SERENDIP team. The Italians are also gearing up for a SETI search, again using a SERENDIP digital receiver.

Nonetheless, it is curious that, despite the widespread interest and belief in alien intelligence, historically rather few nations have actually mounted experiments to track it down. Nearly all the major searches have been made by Americans, Argentineans and scientists in the former Soviet Union, although there have been occasional experiments in

Australia and Holland. Much of the reason for the limited effort is lack of money. Since the collapse of the USSR, SETI projects in that area of the world have stopped, despite Russian enthusiasm and expertise. The research rubles have vanished. Even in America, the entire SETI enterprise involves no more than a few dozen people, and the only funding available is private. But many nations seem to regard SETI not as a matter of dollars, but of sense. The pragmatic attitude of Europeans seems to disincline them to such speculative science. Astronomers in many countries believe in the advice given to Ellie Arroway (played by Jodie Foster) in the movie *Contact,* namely that SETI is a career killer, a field in which a lifetime of toil may reap no interesting data.

Arroway's response to such well-meaning job counseling was to point out that, after all, she could spend her life as she wished. She also insisted that, given the profound consequences of a SETI success, even a long-shot experiment was worth the candle. These were obvious cinematic rebuttals. But those who really do SETI for a living usually cite other incentives. Digital technology is improving at a furious pace, with computer capability doubling every 18 months. SETI receiver technology rides this silicon wave, and the consequence is that today's experiments are always far better than those of the past. Indeed, it has been estimated that Project Phoenix is 100 trillion times as comprehensive as Project Ozma. As the search machinery improves, so do the chances of success.

In addition, despite the many unknowns surrounding the location, abilities, and attitudes of aliens, researchers are able to endlessly imagine new strategies for finding beings that might inhabit the dark depths of the Galaxy. It is a field awash in ideas and promise.

SETI STRATEGIES

The major SETI efforts today fall into two categories: all-sky surveys, and targeted searches of nearby, sun-like stars. The former approach makes no assumptions about where E.T. hangs out, but clearly wastes a lot of telescope time looking at empty space. Furthermore, because each patch of sky is only scrutinized for a few seconds, the sensitivity of such surveys is low. The targeted searches beat this rap by homing in on the Galaxy's better locales. But they usually examine one star at a time. If only one in a million Sun-like stars hosts transmitting extraterrestrials, this strategy could burn a lot of telescope hours before booking success.

Which is the better way to do SETI? The problem is difficult and akin to a hunt for life on a completely unknown continent. Is it faster to carefully sift through a teaspoon of dirt in the hope of finding small, but plentiful insects? Or would success come sooner by flying over the territory scouting for elephants or other megafauna? In this case, it all depends on the relative abundance of big critters versus small. For SETI, the choice of strategies depends on personal preference. Do you think it's reasonable to believe that, among the many societies that could be broadcasting our way, at least a few are doing so with truly powerful transmitters? If so, then it isn't necessary to choose the closest stars. Pointing your telescope at star clusters, nearby galaxies, or even random patches of celestial real estate is the fastest way to find the aliens, since you will thereby sample thousands or even millions of stars at once. Somewhere in this heap of habitats, there will be a "mega-society" pumping gigawatts into its radio transmitters, making itself obvious even from great distance.

On the other hand, it may be that E.T. considers that transmitting such beefy signals is not only expensive, but possibly dangerous. Recall that British astronomer Martin Ryle objected to the three minute Arecibo broadcast of 1974 on the grounds that shouting in the jungle is not a smart thing to do. Maybe the extraterrestrials agree, and keep radio silence except for the long distance communications needs of their own civilizations. In that case, we might not hear them with *any* strategy. After all, present SETI equipment couldn't detect Earth's most common transmissions—either television or radar—from a distance any farther than one light-year. Of course, you might assume that in E.T.'s highly advanced society, every alien will be outfitted with a personal transponder capable of high-definition, two-way television, telephony, e-mail, and remote baby monitoring. E.T.'s planet might be awash in transmitters, so wouldn't all that communication be obvious to us?

Probably not. The trend on our own planet is towards low-power communications. Large, multi-kilowatt transmitters are being supplanted with fiber optics, cable systems, and direct satellite broadcasts. Portable devices, such as cellular phones, radiate very little power into space (this avoids cooking the user's brain, and saves on batteries, too). And if this is the direction that our telecommunications are headed, E.T. is probably already there. His telecommunications may be invisible to us, and SETI will fail to find him.

But there are reasons for believing that this assessment is too somber. To begin with, some easily foreseen activities other than telecommunications would still produce strong signals. For example, aliens might use powerful radars to search for rogue comets that threaten to pulverize their planet. They may have large, orbiting 'powersats': satellites that convert sunlight into electrical energy that is then beamed down at microwave frequencies to the home planet. Such an orbiting power plant would leak prodigious amounts of radio into space. And it is leakage of this type that the limited-sensitivity, all-sky surveys could miss, but the targeted searches might find.

In addition, it's hard to believe that *every* alien civilization is paranoid to the point of not wishing to make contact. This would be like going to a cocktail party at which no one speaks unless spoken to. Under such circumstances, silence would reign. But it never happens, someone always pipes up. Likewise, some civilizations will surely have both the self-confidence and the curiosity to attempt establishing communication with others. They will pipe up with powerful, easily heard transmitters.

Indeed, in the unlikely circumstance that extraterrestrials are exceptionally nearby, we might *provoke* a signal. Within 25 light-years of our solar system lie more than a hundred stars. Any alien occupants around these nearby stars have had enough time to both detect our early FM and TV broadcasts (assuming they have a better receiving system than our present SETI), and transmit a response that we could pick up today. This was, after all, the premise for the signal in *Contact*. Perhaps these neighbors have something pithy to say about our early television programming.

But it's more likely that the nearest civilization is farther, hundreds or possibly thousands of light-years distant. This is much too remote for them to have had the pleasure of listening to our radars or watching "I Love Lucy." They won't know that a technological civilization exists on Earth. Nonetheless, they may have constructed a beacon to signal such civilizations, just in case. This would be a radio analog to a lighthouse, whose powerful beam is offered despite the fact that the lighthouse keeper is unaware of what ships, if any, are in the neighborhood. An alien radio beacon might flash strong signals into space, or slowly rotate to sweep out the nearby territories of the Milky Way. Such beacons are the type of transmitters that our SETI experiments, of whatever variety, could most easily find.

Tilting a radio telescope at nearby, Sun-like star systems or panning one's antenna systematically across the sky remain popular SETI strategies. But there are others. In the late 1970s, two astronomers, T. B. Tang and P. V. Makovetskii, independently concocted a clever idea that might improve the chances of success. They suggested that searches for alien broadcasts could be profitably synchronized to a cosmic event visible to all the Galaxy's inhabitants. The event they suggested was a supernova, the calamitous explosion of a large, dying star. When a supernova goes off, it shines with the brilliance of a hundred billion stars. It's not easy to miss. The ingenious suggestion was that such stellar fireworks could be used as a natural signal flare, telling us when and where to aim our radio telescopes.

This scheme relies on the possibility that an advanced civilization lies somewhere in the galactic territory between us and the supernova. Suppose one does. When the light from the explosion reaches this far-off planet, you can be sure that every alien astronomer will cock his telescope in the direction of the star's blistering blast. But it's conceivable that some members of this extraterrestrial society will be altruistic enough to turn a few of their radio telescopes in the direction opposite

When a supernova is observed by an alien civilization, it may occur to them to broadcast a hailing signal in the opposite direction. In that case, we might pick up the signal after first observing the dying star. Earth, in this representation, is the planet that is farther from the supernova.

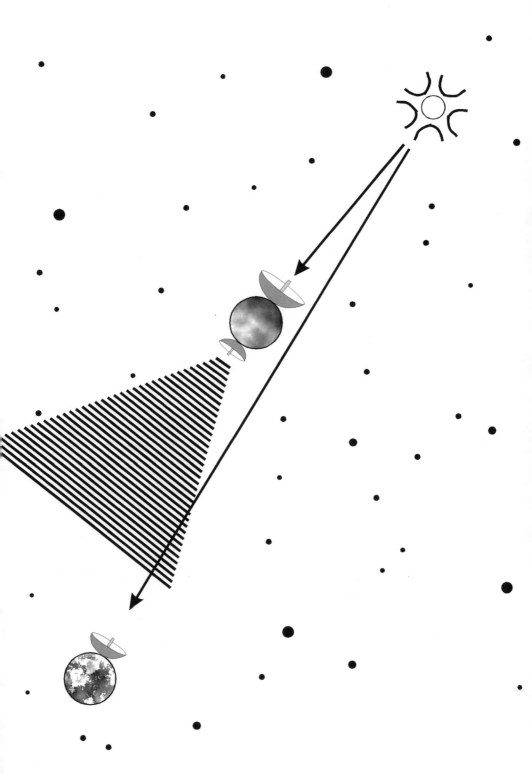

to the big light in the sky, and transmit a follow-up greeting to all those behind. Like a dog chasing a rabbit, their hailing signal would follow the supernova's light through space. Some years later, astronomers located "downstream" would see the exploding star light up their own skies. As a matter of course, they would train their antennas in the direction of this scientifically interesting cosmic show. Shortly thereafter, they would be the surprised recipients of a deliberate broadcas. Expecting to study a rabbit, they would soon encounter the dog.

The suggestion, of course, is that we should be aiming our SETI antennas in the directions of recently-exploded supernovas. As noted, this scheme has the advantage of telling researchers not only where to point their antennas, but when. There are other strategies that do the same. I have proposed that SETI scientists should consider including a special kind of star system in their target lists—eclipsing binaries.

Half of all the Galaxy's stars are found in binaries, that is to say they are members of a stellar pair. Binary stars share a gravitational embrace, orbiting one another as if they were connected by an invisible rod. Most SETI searches shun such double star systems because planets would have a tough time surviving the changing gravitational tug from two suns. The double stars would destabilize the orbits of attendant worlds, even to the point of eventually kicking the planets out of the system altogether. Obviously, if binary star systems are incapable of hosting suitable planets, they won't host extraterrestrials.

But this untenable situation can be avoided if the two stars are separated by great distance, if the "rod" connecting them is a few billion miles or more in length. In that case, planets could contentedly orbit either of the stellar buddies that make up the binary. Imagine that intelligent life has sprung up in such a system. The aliens in a binary would be irresistably tempted to exploit their second, relatively nearby sun. Perhaps it, too, has useful planets. At the very least, it would be an interesting place to build energy-collecting satellites or solar-type observatories. Radio communication links between the two stellar realms of such a binary empire would be eminently practical, with information taking no more than a few hours to cross from one star to another. Radio would be especially effective at reaching all the members of an alien community dispersed among asteroids or in artificial space habitats. Binary empires would generate plenty of radio traffic. That fact alone makes widely-separated binary systems interesting for SETI researchers.

But there's more. Some binary systems are, by chance, oriented edge-on to us. This means that every few years or so we see one star of the binary pass in front of its double. In fact, because these stellar pairs are far away, we can't actually see the two dancing stars. Instead, what we observe is that the light from the system suddenly drops as one sun eclipses the other. However, during the eclipse (which typically lasts a few hours), we are looking along the line connecting the two stars. The rod is pointed our way. If we are fortunate enough to be observing a binary empire, we are at that moment looking down its communication pipeline. Rather than simply hoping that an alien broadcast antenna is aimed in our direction, we can improve the odds by looking down the aliens' throats, as it were. Much like the supernova strategy, observing binary star systems during an eclipse tells us when to expect signals.

Although the schemes described here are, in principle attractive, they suffer from a lack of suitable candidate targets. Supernovae are relatively rare. A large star explodes in the Milky Way only once every three dozen years or so. Eclipsing binaries are not rare, but few are known that are widely-enough separated. Such systems exist in abundance, but since they "eclipse" only once every decade or so, they have been, so far, hard to find.

FUTURE TRENDS

It is nearly four decades since Frank Drake first attempted to eavesdrop on E.T., and SETI has still failed to hear a single peep from the cosmos. One might ask if there's reason to be discouraged.

Those in the SETI business say "No, we've only just begun." The Galaxy is stuffed with a half-trillion stars, and we've sampled fewer than a thousand. E.T. could be broadcasting anywhere on the dial, and SETI experiments have listened in on only a small range of microwave frequencies. Even if we point in the direction of an inhabited stellar

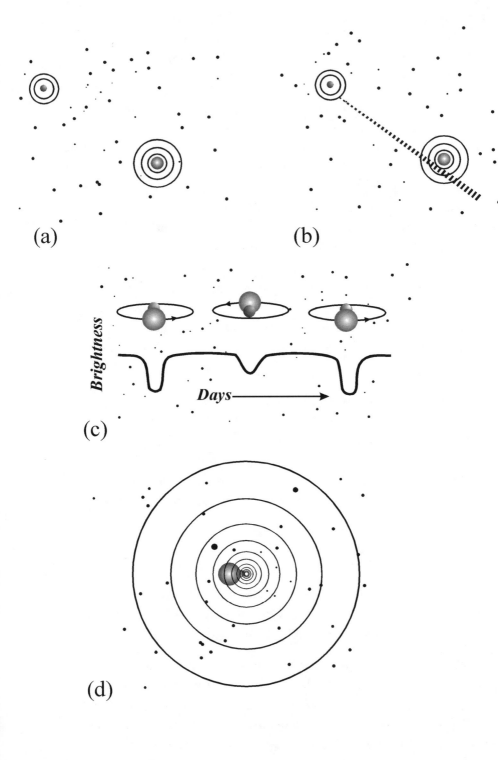

(a)

(b)

Brightness

Days ⟶

(c)

(d)

system, we can't be sure that E.T.'s broadcast antenna is aimed our way at the right moment. Is his transmitter powerful enough? Is it even *turned on?* The standard SETI metaphor is to say that we're searching for a needle in a haystack, and so far we've investigated a teaspoon of hay.

This is certainly true, but it also has a converse. If there is a whole wagon of hay to be had, and we're working our way through it one teaspoon every forty years, then maybe SETI scientists ought to figure out how to get a bigger spoon.

In some sense, they have. Thanks to the incessant march of digital technology, the spoon is always growing. New SETI receivers have more channels. For those researchers busy with all-sky surveys, this means that the heavens can be scanned for signals over a bigger chunk of the radio dial. For the targeted search types, more channels means that stars can be examined faster.

But really major changes in SETI tableware require dramatic moves. Some might be forthcoming. The astronomical community is considering the construction of a new, enormously large radio telescope, with ten times the collecting area of the Arecibo antenna. If built, this new instrument, widely known as the 1KT (for One Kilometer Telescope), would be of such design that it could look in many directions simultaneously. Instead of just one experiment at a time, this telescope could support a dozen. And one of those experiments could be SETI. The 1KT would offer extraordinary sensitivity, and the possibility of 24-hour-a-day searching. It would be dozens of times better than any SETI search underway now.

Another scheme that suggests when and where alien signals might be found takes advantage of a special astronomical circumstance known as an eclipsing binary. About half of all stars are in double systems. (a) If the stars are well separated, they could each host stable planets. (b) In the event that a technological society arises in such a system, it will soon establish communications with its own spacecraft or colonies around the stellar companion. (c) By chance, some double stars will be oriented edge-on to us. These eclipsing binaries are recognized by the fact that light from the system is periodically dimmed as one star passes in front of the other. (d) By observing an edge-on binary at the time of the eclipse, we can look down the communications pipe of the "binary empire," thus greatly increasing the chances that a transmitting antenna is pointed our way.

Researchers are also considering the possibility of looking for something other than radio transmissions. Nobel Prize-winning physicist Charles Townes, inventor of the laser, has often noted that an infrared version of his invention might make an effective interstellar signalling device. Lasers have been built on Earth that can produce extremely short flashes with powers of a million billion watts. E.T. might use similar devices to illuminate nearby star systems with blindingly bright pulses, hoping to get someone's attention. By flashing in the infrared, E.T. could be sure that his laser beam would outshine the light from his own sun, at least over a small region of the spectrum. The idea has merit, and although some preliminary hunts for infrared pulses from nearby stars have been made, comprehensive searches have not. In the early part of the 21st century, they may be.

Other approaches, while ingenious, are less obviously practical. Some folks have suggested that E.T. might be using neutrinos to communicate. If so, the message will be tough to hear, because neutrinos are hard to detect. Others postulate that the aliens are trying to get our attention with gravity waves. But making strong gravity waves requires shaking something as massive as a star. That's a tough engineering challenge. Why would E.T. bother when a table-top radio transmitter could send the same information just as fast?

Then there are those annoying types who are convinced that SETI is missing the boat in a big way. According to them, truly advanced extraterrestrials won't bother with radio, infrared, or even gravity waves to broadcast their greetings, because all such waves travel no faster than the speed of light. New physics, these folks claim, will allow communication that is nearly instantaneous. As in *Contact*, sophisticated aliens will be able to contort space to produce wormholes, shortcuts from one part of the Galaxy to another. Unfortunately, on Earth the only beings that are producing wormholes are worms, and their constructions haven't been of great benefit to interstellar travel. As a very practical matter, there is no way to do SETI experiments with physics we don't yet know. The astronomers conducting the search have a very pragmatic attitude to such speculations: we go with the tools we have. After all, Columbus did. And yes, maybe our technology is naive or inadequate. But one thing's for certain: if we throw up our hands and forego the experiment, we'll never find the aliens.

But what if we do find them? What happens if a SETI experiment succeeds?

The Drake Equation

In November of 1961, the first major gathering devoted to the topic of SETI was held in the burg of its birth, Green Bank, West Virginia. Attendance at this unpublicized meeting, which was an initiative of the Space Science Board of the U.S. National Academy of Sciences, was by invitation only. The ten participants were charged with evaluating the possibility that we might be able to communicate with distant worlds.

Frank Drake, who only a year earlier had shifted the whole subject into high gear with his Project Ozma, tried to organize the meeting's discussion with a simple formula, one that would serve as an agenda for the proceedings. The equation describes the calculation of N, the number of civilizations in our Galaxy that are currently trying to get in touch via radio. The motivation for estimating N is simple: If this number is large, then SETI has a good chance of success. Otherwise SETI researchers may be in for a long wait.

Despite its quantitative cast, the formulation for N, generally known as the Drake Equation, is quite easy to understand. A simple analogy can set the stage for those who feel that any mathematics in this book would be a violation of reader rights, and a threat to their general equanimity. Imagine that the task is to estimate A, the number of Americans who, right now, are seated behind the wheel of their automobiles. One could make a useful, if still approximate, calculation as follows. Begin with the total number of Americans, and then multiply by the fraction that own cars. Multiply this product by the average number of hours each day Americans spend in their motorized buggies, and divide by 24, the number of hours in a day. This product of factors will yield a rough value for A.

The Drake Equation expresses N as the result of a similar multiplication of factors. It has remained the agenda for most SETI discussions, even three dozen years after its formulation. If you wish to impress cocktail party guests with your knowledge of extraterrestrial science, you need only discreetly murmur something about "the Drake Equation."

What values are appropriate for the Drake Equation? N is, as described, the number of galactic civilizations transmitting now. It is the number we want to know. In the following, we'll make some estimates for each of the factors determining N.

$$N = R^* \, fp \, ne \, fl \, fi \, fc \, L$$

In the formula, R^* is the birthrate of suitable, long-lived stars in our Galaxy, in stars per year. A good guess for this number is between 1 and 10. Let's choose 5.

The second term, fp, is the fraction of stars with planets. Recent astronomical work indicates that this number is at least a few percent, and may be as high as 100%. A reasonable estimate might be one-half (50%).

Term three, ne, is the number of planets per solar system that have an environment suitable for life—"earth-like," if you will. Judging by our own solar system,

this number is at least one, and could be higher if Mars turns out to be hiding microbes under its ruddy skin. We choose one as a decent guess.

fl is the fraction of earth-like worlds on which life actually gets underway. On the basis of the rapidity with which biology blossomed on Earth, we can optimistically speculate that this fraction is also one (100%).

Term five, fi, gives the fraction of life-infested worlds upon which intelligence emerges. This is a highly controversial matter, as we've noted. However, we've argued that intelligence has survival value, and convergent evolution will ensure its eventual appearance on any planet where competition for resources takes place. Ergo, we'll be sanguine about sentients, and set fi to one (100%).

fc is the fraction of intelligent civilizations that have the technology and incentive to communicate over interstellar distances. Of course, not every group in a technological civilization will wish to beam radio or light waves into space, but a few can speak for the whole civilization (as television and radar broadcasters do on Earth). So let's set *fc* to one, as well.

The final factor, *L,* is the longevity of a technological civilization, the length of time that E.T. is "on the air." In other words, *L* specifies how many years elapse between a civilization's invention of radio and its self-destruction or disappearance. Before taking a stab at its value, we note that our estimates for all the preceding factors (*R** through *fc*) multiply to 1.0. In that case, the Drake Equation boils down to

$$N = L.$$

So thanks to a convenient choice of values for the first six terms in this equation, we find that the number *N* of transmitting civilizations is equal to the lifetime (in years) of your average alien society. This may strike you as more than convenient. You may view it as optimistic, and wish to contest our estimates for the parameters that precede *L.* For example, while *R**, the birth rate of stars, is fairly well known, the fraction of life-bearing planets that cook up intelligence, *fi,* is quite *unknown*. The values given here are close to what Drake himself chooses, but he is the first to admit that many of

these factors are uncertain. Nonetheless, the lack of precision in determining these parameters pales in comparison to our ignorance of *L,* the lifetime of technology.

L is the term in Drake's equation most dependent on soft science —sociology—rather than astronomy or biology. Additionally it is the one that we know, and perhaps *can* know, least about. In 1971, Carl Sagan said that in trying to evaluate the terms of the Drake Equation "we are faced . . . with very difficult problems of extrapolating from, in some cases, only one example and in the case of *L,* from no examples at all. When we make estimates, we cannot pretend that these values are reliable."

That is a daunting caveat. Efforts to pin down *L* usually amount to guessing what the one technological civilization we know, human society, is likely to do. Within a half-century of inventing radio, we developed atomic weapons, and most estimates of *L* have made the discouraging assumption that we will eventually use them. The majority of these appraisals put the apocalypse early in the next century, in which case it will be an event that many of us will be privileged to experience. If this downbeat scenario applies to E.T., then L is only a few hundred years or so, and the Galaxy is sparsely populated with transmitting societies.

However, it is my opinion that such a view is unrealistically pessimistic. We're not hell-bent for H-bombs here on Earth. This opinion is less a reaction to the diminution of tensions following the end of the Cold War than a recognition of the fact that we're passing through a "bottleneck." There's no denying that we live in dangerous times, perhaps the most dangerous times since the emergence of *Homo sapiens*. Still, it seems unlikely that we could really wipe ourselves out entirely. An example from the past shows how difficult it is to erase human presence. More than half a millennium ago, the Black Plague killed about a third of Europe's inhabitants. This pestilence was the greatest catastrophe in recorded history. Nonetheless, the terrible carnage barely registered as a blip on the growth of world population. In a contemporary vein, various analyses show that nuclear war, and even nuclear winter, would be less than 100% efficient in destroying humanity.

In addition, the danger of obliteration is only temporary. Sometime in the next century, many humans will be off the planet, in artificial space colonies, on the surface of the moon and Mars, or burrowed into the asteroids. Total annihilation of our society will be just as impossible as the total annihilation of Earth's ants. We will be dispersed, and dispersal is the ultimate insurance policy for survival.

This bottleneck—when a civilization has dangerous weapons, but is still confined to a small chunk of real estate—will presumably be encountered by most intelligent, technologically developed species. Since the time scale for getting through the bottleneck is small, only one or two hundred years, it seems probable that many societies will manage to do so. And once past the bottleneck, there's no clear limit. The sharks and turtles have been around for well over 100 million years. Humans might conceivably do as well.

So what's the correct value for L? Frank Drake guesses 10,000 years. But as noted, there's no obvious reason why we couldn't hang out for many millions. And

the same presumably applies to
E.T. So the number of transmitting civilizations N could tally in
the millions.

Without doubt, determining N
is still more sorcery than science.
Drake's own estimate is 10,000
civilizations, considered optimistic by some. But as we've
seen, a far higher number is also
possible. We won't know unless
and until we've found other societies, of course. And we needn't
be discouraged by that hard fact.
Drake's equation was meant to
organize our thinking, and not to
restrict our efforts. Additionally,
if technological societies can travel from star to star, the Drake
Equation may be irrelevant to
SETI's chances of success. If
even one ancient civilization has
spread across the Galaxy, setting
up a massive communications
network to link its worlds, then
a SETI detection might be likely
even if the formalism of this
famous equation leads to a value
for N of only two: them and us.

When E.T. Calls

If it happens, it will begin slowly and without warning in a radio telescope's cramped, cluttered control room. Here, under a hundred tons of steel faced off against the pinpoint gleams of the night sky, a back-burner experiment could change the world.

We can imagine this future drama. The protagonist is a lone astronomer, one of the two dozen or so who have gambled their careers on SETI. For weeks, she has been spending long nights seated in front of a bank of computer displays, nursing a cup of coffee and intermittently scribbling routine entries into a logbook. On the screens, blocks of slowly changing text monitor the electronics that are sifting through the thick cosmic static collected by the telescope. She sips at her drink and scans the displays' laconic reports. There is no theatrical music, no high-tech sound effects; only the constant drone of muffin fans in the electronics racks and the faint, distant grind of the telescope's tracking motor. This is not SETI as depicted by Hollywood. No control room loudspeaker will suddenly break into a squeal or a rhythmic boom. There was a time, many years earlier, when listening for audible signals was practical. Drake did it, after all. But modern SETI experiments monitor tens of millions of channels simultaneously. Computers do the listening.

With only a soft beep as herald, the computers have found that one channel in this multitude bears the hallmarks of extraterrestrial origin.

On the screen, a single line of text tells the tale, a string of numbers giving the signal strength and exact frequency, terminated by the cryptic words "confirmed by FUDD." The FUDD, or Follow Up Detection Device, is a specialized piece of electronics that orchestrates a two-telescope procedure to confirm that a signal is coming from deep space. The astronomer, while taking note, is not excited. After all, the system finds such candidate signals five or six times a week. So far, all have been traced to some sort of satellite interference or other man-made source. None has been extraterrestrial.

Without prompting, the observing software swings the telescope two degrees away from the targeted star system. Ten minutes go by while the receivers accumulate more data. The FUDD then reports that the signal has disappeared, as would be expected if it came from the star system itself. The astronomer pays close attention, but her blood pressure doesn't change. It's probably another satellite, briefly mimicking E.T. as it parades across the sky. Perhaps the signal is from a military jet pilot who, by chance, has just switched off his radar.

The telescope slews back on target. Another ten minutes drag by, and the FUDD reports that the signal has returned. The astronomer puts her coffee down. Her eyes fix upon the display screen. The telescope begins its cycle of on-off observations anew, and the evidence that this signal is extraterrestrial persists. She is now the first witness to a staggering sequence of events, a sequence that has never before occurred. Within hours, she will call another radio observatory to enlist its help. A detection at a distant telescope, by other people and other equipment, will rule out fiendish interference, a bug in the system or an ingenious college prank. Within a few days, the signal will have been confirmed beyond reasonable doubt. The drama begun by a computer's soft beep will have grown to a worldwide din. We will finally have observational proof that other thinking beings populate the Galaxy.

NO SECRETS

Since 1989, there has been a document that lays out what should be done if the scenario described above occurs. Called the "Declaration of Principles Concerning Activities Following the Detection of

Extraterrestrial Intelligence," this short protocol with the jaw-breaker title has been voluntarily adopted by all SETI research groups. Thoroughly uncontroversial, the bulk of the Declaration describes procedures for verifying that the signal in question is truly extraterrestrial. These include a confirming observation at a second observatory, if possible. Once the discoverers are confident that the signal is from aliens and not earthlings, they are directed by the Declaration to make their finding public; to announce the result to the astronomical community, the authorities, and the world at large.

Fine and dandy. But opinion polls show that the majority of Americans not only believe that the extraterrestrials exist, but that they make house calls to our planet. According to the popular view, the government has hidden the evidence for aliens, and maybe even the aliens themselves. Extending the conspiratorial suspicions to SETI, many among the lay public assume that paranoid government agencies would swoop down on the radio observatories to protect the citizenry from possible panic, and hide the news that the extraterrestrials have been found. The scientific discovery of the millennium would be smothered by a government cover-up.

Although it may be a disappointment to those who are certain that the feds will keep news of alien broadcasts under wraps, a cover-up is virtually impossible. To begin with, there is no policy of secrecy and no precautions to ensure it. Even if there were, by the time a detection is confirmed, a hefty number of people will be aware of the discovery, including some at foreign observatories. Finally, the rapid, and occasionally embarrassing spread of information about "close calls" shows that the media tend to err on the side of bravado rather than caution. It's far more likely that a signal will be mistakenly reported as a detection before it's confirmed, than that a *bona fide* result would be covered up.

This is more than just hypothesis. At the February, 1996 meeting of the American Association for the Advancement of Science, a British reporter asked SERENDIP researcher Dan Werthimer whether his SETI team had used their telescope to check out any of the several star systems that had recently been found to have planets. Werthimer said, "Yes, we have. The results are consistent with noise." This standard radio astronomer jargon was misunderstood by the reporter, who assumed that "noise" meant "signal." A story soon appeared in a London paper claiming that the aliens had been heard. Needless to say, they hadn't.

In June of 1997, Project Phoenix scientists felt they might have hit the jackpot when the Green Bank 140-foot telescope turned up signals that passed all the usual, automated verification checks. The telescope was nodded back and forth a half-dozen times, and the signal alternately winked out and returned on cue. After nearly a day's excitement, the signal was traced, not to an alien planet, but to a European solar research satellite, SOHO. This distant bird lies a million miles from Earth and relays information about the Sun back to its European masters with a small, 10-watt transmitter. Despite its low wattage and great remove, the SOHO signal was still strong enough to leak into the telescope and cause SETI scientists to nervously pace the carpet.[8] Remarkably, a dozen hours into this day-long vigil, a *New York Times* reporter called Project Phoenix to inquire about "that interesting signal you're following." One of the office staff had casually remarked on the excitement during a phone conversation to a third party, who then mentioned it to a reporter at the *Times*. An innocuous comment led to the unexpected inquiry. Those who fear that a SETI success would be clammed up greatly underestimate both the urge to share heart-stopping news and the perceptive powers of the press.

Secrecy aside, the orderly release of information on E.T.'s discovery envisioned by the Declaration is probably unrealistic. The process of detection, confirmation, and announcement will be more muddled than methodical. The Internet will be throbbing with rumors disguised as facts. Excited but incomplete reports will wash over radio and television. Supermarket tabloids will go into overdrive, plastering their front pages with exclusive photos of gnarled aliens seated behind microphones. The press conference envisioned by the Declaration may be the first chance the discoverers get to present their story in a

8. SOHO was 90 degrees away from the direction the telescope was pointed, and not at first a suspected interfering source. The signal leaked in via the antenna's "sidelobes," residual patterns of greatly reduced sensitivity in directions other than that in which the instrument is aimed.

straightforward, organized manner. Even so, preconceived notions about E.T. will skew both the way the story appears and how Joe Sixpack and his wife will interpret it.

SOCIETY'S REACTION

If SETI scientists find a signal, some people will challenge the detection, claiming that it's a hoax. After all, there are several million Americans who still doubt astronauts ever walked on the moon. But the reality of the result will be firmly based. Unlike the incessant claims that UFOs are alien spacecraft, belief in a SETI success won't depend on anecdotal evidence or shaky, amateur videos. Anyone with access to a suitable radio antenna would be able to confirm the signal for himself, with his own equipment. There would be no doubt of its legitimacy, and a SETI discovery would immediately precipitate intense scientific inquiry. Every major telescope would be cranked in the direction of ET's signal, in the hope of learning more. But would the discovery also provoke a dramatic response from society at large? Would there be panic? Disbelief? A sudden eruption of brotherly love and international goodwill?

Panic in the streets is unlikely. To be sure, Orson Welles' 1938 broadcast of *The War of the Worlds* did cause a certain amount of alarm. But it's far less threatening to read of a signal from a distant star system than to hear that the aliens are afoot here on Earth, trashing New Jersey. Indeed, if simply knowing that extraterrestrials exist was enough to provoke wide-eyed mobs, then they should already be grid-locking the streets. After all, the majority of Americans are convinced that the aliens are currently in the stratosphere, in government depots, or both.

A better example of the immediate reaction to the news that aliens have been found was the public's response to the August, 1996 announcement by NASA scientists that fossilized Martian microbes had been chipped out of a meteorite. This was, after all, "life in space," even if it was very small and long dead. It was later acclaimed as the biggest science news story of the year, and yet it hardly affected anyone's daily routine. The announcement generated one week of headlines, after which the discovery dropped off the public's radar screen.

The story then became the province of long time-scale media such as books, documentary television, and specialized magazines. There was no panic, and no obvious increase in brotherly love. Philosophical introspection was largely limited to newspaper op-ed columnists. Even the generally vociferous UFO community had little to say; microscopic martians don't qualify as saucer pilots.

Of course, the impact of the idea that Lilliputian life may once have paddled in the ancient seas of Mars was compromised by the uncertainty of the evidence. But the drama of a SETI success would also be weakened by the widespread belief that the extraterrestrials have long ago been sighted. Joe Sixpack will read the news in his morning paper, and think "Aliens? Aren't they already in cold storage in Area 51?" When faced with rigorous evidence of cosmic intelligence, the UFO crowd will defensively claim that they knew it all along. In short, the potential impact of the news of a SETI detection has, in a sense, already been compromised.

No matter how anemic the short-term reaction, most pundits assume that the long-term consequences of finding extraterrestrials will be profound. To prove their point, sociologists have often invoked historical analogs. Copernicus' reorganization of the solar system and Darwin's revision of biology precipitated a "paradigm shift" in our view of how we fit in to the big picture of existence. These events caused us to lose our central role in the physical and biological realms. Demonstrable proof by SETI that we have intelligent cosmic company would surely deliver a roundhouse punch to any remaining hubris, such as the belief that we are intellectually, culturally, or morally superior. This is especially true since detecting one technological society would probably lead to the rapid discovery of more. We would find ourselves surrounded by advanced civilizations.

This would be simultaneously exciting and humbling. But what would be the long-term effect on our philosophy, on our religious beliefs? The answer is unclear. Michael Ashkenazi, of Israel's Ben Gurion University, put this question to a small number of religious authorities from the so-called "Adamist" religions—Christianity, Judaism, and Islam. Adamist religions share a story of man's creation, but also posit a unique relationship between God and man. They hold that we are special. Consequently, you might expect that the discovery of E.T. would threaten that special relationship. Yes, you're the apple of your dad's eye. But how would you feel if you discovered that dad had a few hundred million other apples, nearly all of which were older and shinier than you?

Surprisingly, Ashkenazi says that when he presented the theologians with a scenario in which extraterrestrials had been discovered, "The most common response was amusement." The majority believed in E.T.'s existence, but none felt that this belief caused a problem. Yes, they said, it would be interesting to find the aliens, but their existence wouldn't affect earthly religions. And they claimed the same would be true even if the aliens turned out to be atheist.

Given the historically severe reactions to both the Copernican and Darwinian revolutions, this restrained response might seem a bit too restrained. In fact, it may only reflect a failure to give the matter much thought. There is little doubt that even if mainstream religions respond placidly to the news that we are not alone, smaller groups will find the news apocalyptic. They will see the discovery of cosmic company as heralding a profound shift in the course of history. And they might be right.

But the first few days following a discovery would be more interesting than ominous.

ALIEN MESSAGE

Out of the inevitable chaos following a SETI discovery, the facts of the detection would soon emerge—at least for those who are interested in facts. In view of the known technological limitations of SETI experiments, one can dare to predict what we would learn in those first, exciting days.

To begin with, where is the signal coming from? This might sound as if the answer should be obvious, but it won't be. For SETI searches that scan the heavens, the astronomers who first tune in E.T. will only know that the signal originates from a particular patch of sky. Many star systems could be camped out in such a patch. Even for Project Phoenix, which is a search of specific, close-by suns, a simple detection wouldn't necessarily connect E.T.'s home planet with the target star. By happenstance, a highly remote civilization could lie in the general direction pointed to by the telescope, but be many light-years beyond the target. Fortunately, uncertainty about the source of the signal could be quickly dispelled. Observations with a massively large instrument, such as the 1,000-foot Arecibo dish, could substantially narrow the piece of sky from which the transmission originates. A large constellation of telescopes, such as the Very Large Array in New Mexico (where Ellie Arroway first heard the aliens in the movie *Contact*), could zoom in more precisely on the spot where E.T. is broadcasting.

Once the star system is pinpointed, astronomers with mirror and lens telescopes would check it out, looking for tell-tale stellar wobbles that would betray orbiting worlds. Radio astronomers would monitor any slow variations in E.T.'s signal frequency, slow changes in pitch that would give a first indication of the spin and orbit of his planetary digs.

So within weeks of the discovery, we would likely know a few astronomical facts about E.T.'s home turf. But such information would be, at least in the public's mind, secondary. First and foremost, the world will hanker to know what the aliens have to say. After all, the phone has rung, so what's the message? The public will assume that the SETI Institute's staff cryptographer is laboring into the wee hours, busily decoding E.T.'s transmission.

Alas, there is no staff cryptographer. The popular view of SETI is that it is designed to pick up messages. In fact, as we've noted earlier, the experiments are configured to find steady, or slowly pulsing, narrow-band signals that would pack a lot of energy into a small slice of the radio dial. The counterpart of this type of signal in earthly practice is a so-called carrier wave, a kind of dial tone that underpins most radio and television broadcasts. Carriers have the highest signal-to-noise of any part of a transmission, and would be thousands or millions of times easier to detect than the modulation, or message. But the only information conveyed by the dial tone is that the line is open.

Does this mean we would hear the ring, but not get the call? Not necessarily. MIT physicist Philip Morrison thinks that a great deal of information might be packed into the transmission. It will just come across slowly. "We will be able to read the intention of the senders in the characteristics of the signal itself," he claims. The aliens, after all, are not stupid (at least, none of the aliens we could detect are stupid). They will foresee that only slow variations can be easily found. But even so, according to Morrison, "They won't send an empty carrier. It will have a lot of structure, it will have clues. For after all, it is the purpose of the transmission to inform us."

How might an easily found signal serve that purpose? Perhaps the dial tone will slowly vary. It could be lazily pulsed (like an irregular light-house beacon), and at a speed that would make it simple for us to discover. The meaning of this slow-speed semaphore might be something we could understand. It could be a repetitive message taking years to hear once. Another possibility is that it's merely a pointer, a signal to tell us where to tune to get the rest of the story—the real communication. Of course, these scenarios assume that E.T. has gone to the trouble of sending us a deliberate signal. But as we've noted, intentional transmissions are better grist for SETI's mill than the aliens' domestic radio traffic. Deliberate broadcasts will be more powerful. Aliens that *want* to be heard are more likely to *be* heard.

Clearly, it's probable that our first SETI success will be the result of electronically tripping over a strong (and in hindsight, obvious) carrier wave. But finding the subtle variations of any message will demand a far more sensitive instrument, one whose physical size dwarfs contemporary radio telescopes. We don't have such antennas today; the money to build them is simply not available. However, given the potential importance of E.T.'s interstellar utterances, it seems reasonable to assume that, following a detection, SETI's tiresome battles for funding would be over. Within a few years, a phalanx of gigantic telescopes would be sifting the radio spectrum, looking for the rapid signal variations that would be the bearers of real information. It would finally be time for the cryptographers to get to work, to tackle the ultimate challenge in code breaking: understanding the mysterious wisdom of another thinking species.

It's likely that the cryptographers would have considerable job security, too. Deciphering a message could take a long time. Indeed, it could take forever. As we've mentioned before, any signal we're likely to detect will come from a civilization that is enormously more advanced than we. The aliens' message might be impossible to unravel. Imagine Aristotle's puzzlement if he were faced with the task of decoding a modern color-television signal. And Aristotle was no dummy.

In that case, or in the case that the signal turns out to be merely inter-cepted alien radar devoid of *any* message, we still won't be left empty-handed. We will have proof, after all, of celestial company, we will know that intelligent beings exist. It's just that we won't know much about their particulars. This could be likened to the imaginary situa-tion of the 18th-century South Sea islanders who discover a bottle on their beaches containing a message from Europeans. They can't decode the message, but they know that they are not the only thinking beings around. Their philosophical point of view is affected, but their culture is not grossly altered.

A more interesting scenario is that they might try to help us. A delib-erate transmission might be *designed* for infant technologies such as our own. In that case the alien senders might take pity on our cryptog-raphers by sending simple directions for deciphering their broadcast, a kind of interspecies primer. Elementary mathematics is frequently sug-gested as a good first lesson. Indeed, the initial messages received in the movie *Contact* amounted to a bit of boring algebra. On the other hand, E.T. may figure that anyone able to eavesdrop on his signals will probably have already taken algebra, and send us pictures instead.

 Pictures, made up of simple pixel arrays, are easy to decode, and can contain a lot of intriguing information ranging from "Here's what we look like" to "Build this machine!" The former message is one we would obviously find interesting. As described earlier, we've been nar-cissistic enough to send our own appearance into space with the 1974 Arecibo radio broadcast, and on the Voyager and Pioneer records and plaques. The "Build this machine" message was transmitted to Earth by the aliens in *Contact,* who urged us to construct a wormhole gener-ator that would allow Ellie Arroway to make some quick trips to other galactic haunts. While plausible, such alien-inspired construction projects seem more than a little challenging. Even if we were sent details for a machine, it's not clear we could actually put the thing together. Fabricating anything complicated requires a lot of technical infrastructure. Imagine providing Neanderthals with plans for a personal computer. Even if the instructions were first rate, the Neanderthals wouldn't be able to build it (and even if they did, they would probably get hung up on the operating system). Nonetheless, given enough time and patience on the part of our extraterrestrial tutors, there's no certain limit to what they could teach us. Consequently, if we can crack the code of an alien signal, with or without E.T.'s help, the impact on earthly society could be profound. We would be in touch with an ancient and sophisticated culture. We might skip eons of history and leap into what otherwise would be a far distant future.

Some consequences of this sort of contact are obvious. We could bone up on advanced physics, chemistry and astronomy. The aliens might fill us in on how to get along. Perhaps they would even be considerate enough to divulge the cure for death (a good thing to know from the individual's point of view, although a real challenge for society. How would civilization avoid suffocation under a crush of immortal human protoplasm?) At the very least, the aliens' existence would constitute proof that a culture can survive its own technology. There would be cause for optimism on Earth, for if E.T.'s society could endure for countless centuries without self-destructing, then presumably ours can as well.

While far from assured, all these benefits are possible. On the other hand, some researchers who have dared to consider the possible consequences of a detection are less upbeat about what happens when E.T. phones Earth. Maybe the aliens would just wish to exploit us. This suggestion has considerable heft, simply because history is filled with examples of exploitation on Earth. A radio signal might not seem to pose the same degree of threat as that which confronted the Incas when the Spaniards ambled into town (or—the modern equivalent —when the aliens in *War of the Worlds* invaded). But what if E.T. patiently taught us how to build innocuous-appearing machines that, instead of transporting us to his stellar home, surprised us by either producing aggressive, proxy robots, or by chemically cloning the extraterrestrials themselves? Such an outcome could make us unwitting hosts to an alien intelligence that uses earthlings, and presumably other beings, to aid their spread throughout the Galaxy.

This smacks of dark science fiction. But more credible is the one-sided damage that could occur were we suddenly presented with information far beyond what we know. How important would our day-to-day activities seem to us if we were suddenly privy to the workings of a truly sophisticated culture? Would an alchemist of the twelfth century have continued to toil in his laboratory if suddenly faced with the chemical knowledge of the twentieth? How motivated would you be to punch the company time clock if the wisdom of a thousand millennia were washing over Earth? We might revert to a child-like state and believe that only our extraterrestrial superiors are capable of doing anything that's significant. Sebastian von Hoerner noted that "if a Stone Age culture comes into contact with us, this means absolutely the end of that Stone Age culture, although this may take a while; and if we come in contact with some superior civilization, this again would mean the end of our civilization, although also that might take a while. Our period of culture would be finished and we would merge into a larger interstellar culture."

This fate, if indeed it is our fate, might conceivably be a good thing in the long run. Consider the example of Japan, whose encounter with foreign cultures is in stark contrast to the experience of the Incas. For two hundred years, Japan deliberately isolated itself from the West with a policy of *sakoku*. But shortly after the American Admiral Perry opened up the country in the 1860s, this island nation decided to join the rest of the world. Japan not only adopted Western technology, but incorporated much Western culture into its own. Its success has been so impressive that it occasionally discomfits its one-time mentors. Those who fear SETI efforts because of the possibility that it would put us into actual cultural contact with aliens, who insist on isolation for Earth, may be advocating the same mistaken policy adopted by Japanese emperors of the early 16th century.

Enforced ignorance is not bliss. But such thoughts naturally raise the question of whether we should deliberately encourage interplanetary pedagogy by letting the aliens know that we got their message. If we hear their signal, should we answer? Transmitting a reply may seem superfluous in light of the torrent of television and radar signals that have been boiling off our planet for a half-century. But a directed response would not only be far stronger than this unintentional leakage, it would be at the frequency of their signal to us—a frequency the aliens will presumably monitor. So broadcasting an answer might make sense if our intention is to join an interstellar rag chew.

However, a reply from Earth raises some difficult political and social questions. Who will decide what to say, or indeed if anything should be said at all? John Billingham, now at the SETI Institute, is attempting to tackle this thorny, if still hypothetical problem. He has long headed a committee of the International Academy of Astronautics that wishes to extend the Declaration to treat the matter of deliberately transmitting to E.T. "Some people think at this stage it's not worth indulging in such thinking," Billingham admits. "But the significance of transmitting from Earth is potentially so profound in its implications for our civilization that we should consult extensively at the international level beforehand." The principles espoused by Billingham's committee are that any message sent to our cosmic neighbors should reflect the will and wisdom of all Earth's peoples.

This is the United Nations approach. It resonates with admirable idealism, and it also deals with a potentially awkward situation in which third-world countries without large radio telescopes could feel cut off from developments important to all mankind. But reality would probably short-circuit these proposed diplomatic niceties. While the UN members were trying to reach consensus on what exactly our message to E.T. should be, the populace will take matters into its own hands. Just about everyone with a backyard satellite dish will swing it in the direction of E.T.'s star system. Those with the ability to cobble together a transmitter will be broadcasting their own personal philosophies. Radical political and religious groups will quickly assemble radio hardware in an attempt to bend E.T.'s small ear with their pithy polemics. Physicist Freeman Dyson has said that, following a SETI success, there will be a mad rush to get on the cosmic airwaves. Earth will reply in a very human way: with cacophony.

Concerns about who should reply or what they should say may be moot anyhow. E.T.'s home could be a thousand light-years distant or more. If so, there's no real hurry to grab the microphone. Conversation is going to be excruciatingly slow, no matter how fast we respond.

This raises an interesting point. If E.T. is distant, his message to us is old, having left his planet centuries or possibly millennia ago. Of course, just because the program is dated doesn't make the broadcast less interesting. We happily, and with profit, read the writings of Aristotle or Caesar, even though their messages were sent a few thousand years ago. And in such cases, there's no chance for a reply. The senders are long dead. Might the same be true of E.T? Could his civilization have disappeared in a radioactive puff sometime between transmission and receipt? While possible, it's unlikely that a long-lived technological civilization of the type we expect to hear will have the bad form to self-destruct before we get in touch. It's as unlikely as the possibility that Aunt Clara will keel over between the time of writing you a letter and its delivery to your home. And even if the aliens do conk out, that might not matter in the long run either. If we find one civilization, we will undoubtedly soon find others. After all, we would now have some idea of frequency and signal strengths used for interstellar transmissions. In short order, we would likely uncover many broadcasting societies, establishing ourselves as simply one civilization in a far-flung community—a community in which members are already in touch. We will be the new kids on the block, and we will have much to learn.

CONSERVATIVE APPROACH

Most people believe we share the universe with intelligent aliens, at least people in countries where pollsters have bothered to ask. But their view of extraterrestrials is strongly biased by the anthropomorphic, understandable aliens of cinema and television. Even astronomers are in the thrall of this conservative assumption that "they're like us," in behavior, construction, and motivation, if not in appearance. We've looked at life's origin and evolution on this planet, found the process comprehensible (if not fully comprehended), and extrapolated it to E.T. The discovery of other planetary systems and the suggestive evidence for ancient life on Mars has encouraged us in the view that we are, indeed, biologically "mediocre."

Of course, this is still a premise without proof. We've pursued its consequences in this book not so much because it predicts an inhabited universe (although it does), but because it gives us some idea of how to find co-inhabitants and what to expect of them. It is a baseline approach, producing a conservative set of predictions. It suggests that billions of other earth-like worlds, still unseen, silently twirl in the Galaxy's distant stellar realms. Some of those worlds would have long ago produced life, and some of those, presumably, now host thinking beings.

But SETI, if it succeeds, may show that we've been entirely *too* conservative. After all, the habitats of the Galaxy are varied, vast, and dispersed. The inhabitants may be wildly diverse as well. Yes, planets are attractive for life, and possibly the only locales where biology will begin. But life need not forever be restricted to its native world. Dispersal, at least within a solar system, is not formidably difficult. As the Martian meteorite ALH 84001 hinted, even bacteria can book passage to nearby planets. Humans and other intelligent beings can manage such other-worldly emigrations for sure.

Indeed, there is an obvious way such space pioneering could get underway. We've noted that population pressure on Earth may soon incite us to build spinning, artificial space stations—massive aluminum cans that would provide the good life to a few million earthlings each. These canned colonies wouldn't need to orbit our planet, of course. They could loiter in the vicinity of the moon, or between here and Venus. It's not hard to envision a future in which the overwhelming majority of humans live off-Earth: on the moon, Mars and possibly the satellites of Jupiter and Saturn, in artificial space colonies, and on the thousands of asteroids that pepper the solar system. While the idea of living in an oversized metal can or a colonized hunk of rock may seem fourth-rate, the facts will surely prove otherwise. The climate will be ideal, the views will be spectacular, and the mosquitoes will be non-existent. We will undoubtedly do this, and we can assume that E.T. already has. Additionally, as we've noted, dispersal is good insurance against calamity. While large-scale war or natural disasters (such as an errant comet) could possibly wipe out a single planet's inhabitants, it's inconceivable that they could snuff out every member of a dispersed community. Moving off its home turf will give an alien society a better chance at long-term survival.

That's good for them, and good for SETI. After all, if sophisticated societies routinely self-destruct within a few centuries of inventing radio, we'll never hear them. Their transmissions will be like flashbulbs randomly popping off in the dark of night. It's unlikely that we'll be looking in their direction when the light is on. Only long-lived societies can transmit long-lived radio beacons. The civilizations we are likely to hear will be long-lived. They will be societies with thousands or millions of years of technology under their communicator belts. They will have dispersed. Consequently, it's an excellent bet that any detected aliens will no longer be restricted to the planet that produced their earliest ancestors.

But exactly how dispersed might they be? Must they endlessly hang out around their native sun, or could they escape their solar system entirely? We've discussed the difficulties of interstellar travel, in particular the long travel times. But a canned colony could, in principle, bolt a rocket engine to one end of its habitat, and slowly wend its way to another star. To do so would be a challenge, but not impossible. And to the extent such *interstellar* dispersal occurs, we can expect to find advanced aliens around stars that might be patently unsuitable for incubating life.

It's possible that, from E.T.'s point of view, "no sun is a bad sun." But we should also consider the chance that alien intelligence could be found far from *any* sun, camped out in the true deserts of the Galaxy, the space between the stars. Might aliens exist in these empty realms? The obvious impediments to true space colonization are the lack of raw materials and energy—only the faint glow of distant stars breaks the everlasting blackness of interstellar space. Since we imagine E.T. to be a member of a large society, such galactic wastelands are unappealing. They don't have the matter and energy to support huge, concentrated populations.

But what if E.T. isn't biological? A machine needn't be composed of billions of more-or-less identical entities, the way human society is. The capabilities of artificial intelligence wouldn't be limited by how much processing power could be crammed into a cranium or thrust through a birth canal. One good machine could, in principle, do just the sort of cerebral things a large number of creatures could do. It wouldn't require a massive infrastructure of farmers, taxi drivers, or lawyers to keep itself operational. A single device could be both intelligent and self-sufficient, given a modicum of energy and occasional supplies of matter for parts replacement or hardware add-ons. Compared to a biological society composed of a few billion creatures, these demands are modest.

So it's conceivable that even the otherwise dismal spaces between stars have their occupants. "Loner" machines might ply the galactic deserts. The more ambitious of these would search out oases—dense, interstellar clouds of gas where raw materials can be found in greater abundance, and where newborn stars gush energetic light into space. Such locations are too young to have incubated life, but they might attract itinerant machines questing for rare, heavy metals in ore-sized chunks.

What would be the purpose of these solo sentients? Can we expect that they would ever betray their presence with a radio signal? Or would they choose to remain cryptic and unseen, to avoid competitors or destructive interference? We can no better guess their motivations than goldfish can infer ours, but given the daunting distances of the Galaxy, they might spawn a few probes to discover and reconnoiter new habitats. This would necessitate machine-to-machine communication, producing radio traffic that we could chance to overhear. Once again, this type of scenario suggests that SETI scientists consider aiming their radio telescopes at some unconventional targets—the giant molecular clouds, and the dense, central regions of the Galaxy. But it also hints at the manifold opportunities that the universe offers for intelligent "beings"—whether biological or machine. Given that these habitats have existed for roughly ten billion years, it would be shortsighted to write them off simply because, at first blush, they are unappealing. When we swing our radio telescopes towards the heavens, we are looking for intelligence, after all, not biology.

That we share the universe is still a belief, not a proven fact. But unlike those distant times when the ancient Greeks imagined that all the lights of the sky were populated by gods or men, science now bolsters belief. The delicate radio footprints proving that others are out there might be found tomorrow, or next year. What these faint signals will tell us can only be guessed of course, but it's unlikely that the aliens' message will be as straightforward as movies and television suggest. The extraterrestrials in our stories are, for dramatic reasons, similar to us in motivation and intelligence (even if these strange beings often choose to emulate our more destructive tendencies). They are not so 'alien' that they can't be understood in the context of human experience. Consequently, comprehending the yet-to-be-heard extraterrestrials that SETI seeks is perceived as no more challenge than understanding someone from an obscure part of our own planet.

But as the English biologist J.B.S. Haldane observed, "the universe is not only queerer than we suppose, it is queerer than we *can* suppose." His oft-cited remark applies admirably to modern astronomical research. We now catalog a host of cosmic objects—black holes, pulsars, quasars—that were unimagined as the 20th century began. Those who expect the aliens, and their effect on humankind, to be less than extraordinary should take heed.

INDEX